Neil Armstrong

The Lives and Careers of the First Men on the Moon

(The Life and Legacy of the First Astronaut to Walk on the Moon)

Kevin Okelly

Published By **Phil Dawson**

Kevin Okelly

All Rights Reserved

Neil Armstrong: The Lives and Careers of the First Men on the Moon (The Life and Legacy of the First Astronaut to Walk on the Moon)

ISBN 978-1-77485-662-8

No part of this guidebook shall be reproduced in any form without permission in writing from the publisher except in the case of brief quotations embodied in critical articles or reviews.

Legal & Disclaimer

The information contained in this ebook is not designed to replace or take the place of any form of medicine or professional medical advice. The information in this ebook has been provided for educational & entertainment purposes only.

The information contained in this book has been compiled from sources deemed reliable, and it is accurate to the best of the Author's knowledge; however, the Author cannot guarantee its accuracy and validity and cannot be held liable for any errors or omissions. Changes are periodically made to this book. You must consult your doctor or get professional medical advice before using any of the suggested remedies, techniques, or information in this book.

Upon using the information contained in this book, you agree to hold harmless the Author from and against any damages, costs, and expenses, including any legal fees potentially resulting from the application of any of the information provided by this guide. This disclaimer applies to any damages or injury caused by the use and application, whether directly or indirectly, of any advice or information presented, whether for breach of contract, tort, negligence, personal injury, criminal intent, or under any other cause of action.

You agree to accept all risks of using the information presented inside this book. You need to consult a professional medical practitioner in order to ensure you are both able and healthy enough to participate in this program.

Table Of Contents

Chapter 1: A Ramblin' Man 1

Chapter 2: Armstrong At Edwards Air Force Base On 1956 19

Chapter 3: Slayton 43

Chapter 4: Armstrong Prepares For Gemini8 64

Chapter 1: A Ramblin' Man

Neil Armstrong

Neil Alden Armstrong (born August 5, 1930) was born in Wapakoneta to his maternal grandparents. Stephen Koenig Armstrong was working back home as an auditor when Mrs. Engel contacted him to tell him that Viola, his wife, had given way to Neil. Baby weighed 8 3/4 lb. Viola was very sick between 9:30 P.M. and 12:31 A.M. Doctor had to treat it with instruments. Viola seems fine but is weak and sore. Baby is very cute. Viola won't be hurt. If Viola is having a bad day, we'll let her know. Tomorrow, more to come. Viola wrote her love to Neil." Stephen reached out quickly to Viola, little Neil and drove them to Warren in Ohio. Stephen's career required the family to move around often so they didn't stay for too long.

Stephen loved his son, and often took him on outings. Neil had a younger sister, June, who was now living in the family's cradle. Neil's older brother Dean was born in 1935. June was the best friend and playmate of Neil's childhood. June later complained well-naturedly that Neil wasn't there when it was time to plant potatoes at their grandparents' farm. He would usually be in the living room, reading a novel, and would often be found in the corner. Also, I don't think everyone was allowed to put [model aircraft] glue all over anything. But he was. He never did any wrong. He was Mr. Goody Two Shoes, if one was ever created. It was simply his nature."

Neil wasn't always a "goody two-shoes" guy. Even his father, Viola, wasn't as committed to Christ as Neil. Neil's very first airplane ride occurred right before his 6th Birthday. Stephen later told Stephen that the plane ride to Warren airport was more affordable in the morning than the price at lunch, which he later claimed was the reason he missed Sunday school. So, Sunday school was skipped and we took our first flight. ... Old Ford Trimotors. They rattled. Neil was thrilled and scared of me."

Neil loved to build model planes. But, unlike most boys his age who fly them, he didn't often fly his. He explained, "I hung my models with string from my bedroom ceiling. I put a lot into them and didn't want to crash them. It was rare when I flew one. He noted, "You cannot have success with a model that wasn't built well. My focus was more in the building than on the flying." Even though I was still in elementary, my intention was to become an airplane designer. I chose to become a pilot after realizing that a good designer needs to be proficient in the operations of an airplane. I read many aviation magazines like Flight and Air Trails and Model Airplane News.

After moving 16 different times in 14 year, the Armstrong family settled down in Wapakoneta. Neil was enrolled at Blume High School. His studies were exceptional and he showed particular aptitude in the sciences. He also took flight lessons at Wapakoneta's nearby airfield. Because of this, he received a pilot's license before he received his driver's license.

Armstrong was an Eagle Scout as a result of his upbringing. According to Kotcho Tolacoff, one scout in Armstrong's troop who participated in this program, the young men were required to hike 20 miles. They set out one morning to find Carey. After finding it, they found it 10 miles away. But, upon reaching their destination, they decided that lunch would be a good idea. Armstrong had to get to work and he drove the pace. Despite feeling fatigued, Solacoff remembered how Neil kept pushing him to push himself to do more so he could get work done. We told Armstrong that he should go ahead. Armstrong then began walking and running in a pattern known as "the Boy Scout pace".

Armstrong, an accomplished individual who entered Purdue University right after his 17th year, majored at aeronautical engineering. The Holloway Plan provided financial support for his studies. It was a bill "to provide training for officers for naval service and other purposes" that paid for his tuition. He would then complete his bachelor's program at Purdue with two years of flight and another year as Naval Aviator after completing two years.

Armstrong was not able to join the Naval Reserve Officers Training Corps. He also did not enroll in any classes related to naval science. Instead, he joined Aeromodelers Club. His activities included: "I would win events or get 2nd place ...[racing an] gasoline-powered control-line model, flown wires, operated at the centre of a ....circle." I learned a lot from people and gained much more insight and experience, some of which were World War II veterans.

It's ironic, given Armstrong's fame in a field few had imagined, that Armstrong felt that he had missed out on the most exciting time of flight. Armstrong later complained that the times had changed "By my time I was old enough and became pilot, everything had changed." The great planes I had admired growing up were disappearing. I had grown up admiring the chivalry and bravery of World War I pilots Frank Luke. Eddie Rickenbacker. Manfred von Richthofen. Billy Bishop. However, by

World War II aerial chivalry appeared to have diminished Air warfare was becoming less personal. The record-setting flights--...Lindbergh, Earhart, and...--across the oceans, over the poles, and to the corners of Earth, had all been accomplished. I felt resentful. I was disappointed to learn that my generation was late in history, despite being passionate about and fascinated with flight. I had missed all my great adventures and moments in flight."

Armstrong began to doubt his faith, as did many young men from Christian homes. Armstrong maintained a faith in God and did not forget his mother's Reformed Presbyterian teachings. This was something that upset his mother for the rest her life. In 1969, she stated, "Neil was an enjoyable child for us to raise in all ways, but when was a senior in High School, and even more in College, he began questioning the truth of Jesus

Christ." I felt that he was not praying ..... I can see how his whole life has co-incided. God gave him the ability to think and maybe even destined him to the job he has been doing. As a boy and as an adult, he loved God's creation and was fascinated by it. It was almost as if he was being called by the heavens--so great his insatiable interest. He has been good and smart, a scholar and thinker, as well as a hard worker. I've listened to every one of his speeches very carefully. Although I am not an expert speaker, I feel every word was thoughtful and from the heart. His thinking is broad and his thoughts far-reaching. He seems to be inspired in God's direction and speaking Gods will. For this, I am immense grateful."

Armstrong did not doubt his faith in God. However, he never doubted his mother's love and that of his entire family. Armstrong wrote letters often. While only a few of Armstrong's letters to his relatives survived, the ones that survived were full of warmth and details about his school days. He wrote "Thanks" in one of them: "Thanks so much for the laundry, letters and Girl Scout Cookies. They were nearly gone when I returned from work last night. It doesn't matter if I get a job. I am going to school for summer. It's a must. I have my class schedule prepared. It is: Differential Calculus. 8-10 A.M....Physics 11-12 A.M....Physics Lab 13-15 P.M. "Today we went to Indianapolis to participate in the first model airplane contest. My control lines fell on the first official flight. This prevented me from having a chance at winning anything. I think my studies are improving lately. I like analytics and I understand a bit of the Chemistry we are learning." He stated, "I saw tonight a show that was the most impressive I had seen in years. It's called Sitting Pretty with Clifton Webb Maureen

O'Hara & Robert Young. It is recommended especially to you and Dad. It is hilarious.

Skiing

Neil Armstrong: "The one thing I regret most was that my work required so much of my time, and a lot traveling."

Armstrong received his notice to report on January 26, 1949 to Pensacola Naval Air Station. Here he would train as a midshipman and was promoted. Armstrong was a North American SNJ Trainee, eventually flying solo on September 9, but he also trained in a North American SNJ Trainer. On March 2, 1950, he landed his first flight on an aircraft carrier, USS Cabot.

The USS Cabot

Armstrong left Pensacola for the Naval Air Station Corpus Christi in Texas to learn how to fly a Grumman F8F Bearcat. Eventually, he had to land onboard the USS Wright. He described the experience by saying: "I asked for fighters and was fortunate to be assigned to fighters. Fighter pilots used to say that only the most skilled men could become fighter pilots. My own belief is that a lot of it was related to what the navy had at the moment you graduated. Because in my particular class, most people got what their asked for. People from other generations can remember saying no one got what their asked for. ... The F8F-1 Bearcat was given to me as my advanced trainer aircraft. It was an extremely high performance aircraft that I was happy with.

Armstrong discovered that he was officially a navy aviator just weeks later, on August 16, 1950. From there, he transferred to Fleet Aircraft Service Squadron 7(FASRON 7) in San Diego Naval Air Station. It was here that he became the youngest member of the all-jet squadron VF-51. He flew his first aircraft, a Grumman F9F Panther in less than six days, on January 5, 1951. This success earned him the Navy's promotion to Ensign on June 5. Two days before his first jet landing on Essex, he was also promoted to Captain. Three weeks later, he sailed for Korea with the Essex.

F9F-2 Panthers over Korea

Armstrong became a well-known veteran flier quickly. His life was almost ended when he flew a mission on September 3, 1951. According to Navy's official account, "Ensign Neil ARMSTRONG VF-51 saved himself with a little bit of exceptional headwork." He was attacking a target high up in the hills. He was still in his run when he got hit by AA. He lost elevator controls, but in less than a second he was able roll in all the back tabs he could. His plane, loaded with ordnance, was so close it crashed to the ground that his pilot cut two feet off the starboard side of his wing with a power pole. He was able, after adjusting the stick to the trim tabs, to fly to friendly terrain and safety. A plane with stall characteristics exceeding 170 knots required positive elevator control to land. The bailout was therefore necessary. This was the Air Group Five pilot's first ejection chair bailout. ARMSTRONG ejected his self, cleared the seat, opened the chute and landed near K-3 with no further incident." Armstrong jokingly stated, "Twenty foot from Mother Earth at this speed is awful doggone far!"

Armstrong's Korean Was war service was short and uneventful, but it brought him many merits. He flew an incredible 78 missions. This gave him 121 hours. Before he was done, Armstrong received numerous awards and certificates, including two gold star, the Korean Service Medal & Engagement Star, the National Defense Service Medal and United Nations Korea Medal. He thought the Air Medal was the best. It had this glowing citation attached: "For distinguishing him by meritorious achievements in aerial flight as a Pilot of a Jet Fighter in Fighter Squadron FINTY ONE, attached at the U.S.S. ESSEX, during attacks on hostile North Korean Communist Forces. Between 21 August 1951 - 9 October 1951, Ensign Armstrong participated, despite grave dangers, in 20 flights that included strikes on transportation and lines to communication at HAMHUNG. MAJONNI. PUKCHONG. And SONGJIN. He accomplished his assigned missions with skillfulness and courage. His devotion was in full compliance with the United States Naval Service's highest traditions.

Armstrong was released on May 9, 1952. Armstrong continued to serve in the Navy Reserve. His rank as a lieutenant, junior grade, was his. This enabled him to continue flying, first in Glenview Illinois with VF-724 and then later in Los Alamitos California with VF-773.

Armstrong in 1952

Armstrong was still a war veteran, and he needed to finish college. Armstrong returned to Purdue for his degree in 1952. Armstrong rose in grade, held the position of chairman of Purdue Aero Flying Club, pledged Phi Delta Theta, played baritone and was part the school's marching orchestra. Before graduating from Purdue in January 1955, he had also written and directed a stage version Walt Disney's Snow White and the Seven Dwarves. Janet Elizabeth Shearon (a Home Ec major) was engaged to him at that time. Janet describes their relationship's origins as: "Neil known me for 3 years before he ever asked for a date. He told me later that Neil had told his roommate the first time he saw me was when he returned home from work to inform him that I was the woman he was going on marrying. Neil isn't one for hurrying into anything." Neil and Mary were married January 28, 1956. Neil then moved to Los Angeles. Neil had been assigned as a High-Speed Flight Station Officer at Edwards Air Force Base by the

National Advisory Committee for Aeronautics.

Chapter 2: Armstrong at Edwards Air Force Base on 1956

Armstrong established himself as a legend both on the plane and at the ground. Armstrong was a mechanical engineer graduate, but it proved difficult to maintain his own car. Milton Thompson was also a test pilot. He said that Neil lived in an apartment on a couple of acre in Juniper Hills at the Antelope Vale. Neil owned several cars, and none of them had good mechanical condition. Neil devised a method to remedy the questionable condition his cars. His home was situated in the hills above Pearblossom Road. On the way to Edwards, he would simply begin rolling down the hill on one of his cars. If the car started to run and his car sounded okay, he would continue along Pearblossom Hwy and head for Edwards. He would take a left at the highway, and drive down to an automotive shop if it didn't work or it sounded bad. He would then turn around and walk up again to test another car.

Armstrong spent most his time at Edwards in chase planes to test bombers. After that, he flew the bombers. Armstrong's first adventure at Edwards came during one of those missions. Gray Creech, Gray's biographer, claims that Armstrong was riding in a Boeing B-9 Superfortress to drop the Douglas D-558-2 Skyrocket. When something went wrong, Neil Armstrong, who was pilot, saw the propeller hub appear to shoot past the cockpit. He looked at the number 4 propeller and realized it had broken. Armstrong and Stan Butchart, pilot, coolly reacted, and tested the bomber's controls. Butcharts had gone, but Armstrong maintained some flight control linkage. They prepared the aircraft for an Emergency Landing. They had been trying unsuccessfully to feather their number four propeller for some time. They had already jettisoned the D-558II Skyrocket with pilot Jack McKay aboard. The Skyrocket's stuck valve meant that they couldn't land on time. Also, the propeller problem created a large workload. McKay was able to safely land the Skyrocket onto the dry lakebed below.

Armstrong and Butchart were able to land their craft safely, and Armstrong then went on as the project pilot for the Century Series fighters. He also flew various Lockheed and Douglas aircraft. Parasev also included him in its paraglider research program. He flew more 200 different types of aircraft, and might still be an aviator if not for the new and exciting opportunities.

NASA and Space Race

Janet was pregnant by Eric in 1957. After giving birth, she quit school to become a wife and mother full-time. Karen was conceived on April 13, 1959. While the family had a happy childhood, tragedy struck in June 1961 when Karen received a diagnosis of a malignant brain cancer. It was then that happiness was replaced in a constant state of grief and stress. Grace Walker (a friend of Neil's and an astronaut) described meeting Neil and his child shortly after diagnosis. They wanted the opportunity to see our daughter [a September 1961 born girl]. I got Karen out of her bassinet, put her on the beds and Neil carried Karen. She could touch her and sort of hold her. Karen was a determined little girl. I wanted it to be more public, but Neil did not consider it acceptable. For example, I would have put my hands on Karen or said a prayer. Neil came to my aid because he wanted to encourage Karen. It was evident that he cared deeply for his little girl.

Walker could empathize with their grief, since Joe, her husband had just lost a son. Armstrong struggled to grieve. She said, "I would say that it is a pilot thing." They're usually quite content. They would say that they had an 'ok flight', and then they would go to the bathroom and vomit. Joe was much more supportive of Janet than Neil. But I'm not saying that to be a criticism. Just the way Neil was, he was very close emotionally." Janet lost her mother Karen on January 28, 1962. Neil could not comfort Janet because he was so focused on his work that he didn't want to feel the pain. Walker observed that Neil often used work as an excuse. He did his best to keep away from the emotions. I know that he felt terrible for Karen. He was simply trying to get over it. Jan was angry for a very, very long time. I was angry at God and Neil.

Armstrong's future was changing. He flew the Bell X-1B (his first rocket-powered ride) to speeds exceeding 1,000 miles per hour at an altitude over 11 miles. While the landing gear on his nose failed to deploy at the time he landed the rocket, he was able to make it safely out. The new

National Aeronautics and Space Administration, which he joined on October 1 1958, was already a fascinating space program.

A Bell X-1B

Neil flew the MH-96 he tested to 207,000ft, a height not reached until the Gemini 8 Program. To illustrate a flaw of the plane's glide limit, he held his nose up during the descent. It produced shocking results. According to the flight record, Armstrong maintained a G-limit for reentry. This caused Armstrong's aircraft to bounce back from the atmosphere. After traveling approximately 45-five miles to the south of Edwards, the X-15 turned around. Armstrong's first flight over 200,000 feet. Armstrong's highest altitude reached and his longest flight. "Longest duration flight in program, 12.4 min." This report did not mention that Armstrong barely missed a Joshua Tree stand when he finally landed.

Armstrong and an X-15

Tom Wolfe, who wrote The Right Stuff about the Space Race's early years, said that Armstrong's expression had hardly changed. If you asked Armstrong a question, his eyes would be pale blue and he would continue to stare at the questions. After figuring out that you didn't understand his answer, you would click on his lips and hear a series of very long sentences. These sentences were filled with anisotropic effects and multiple encounter trajectories. . . . It was almost like his hesitations were just data input intervals for his computer. This description was accurate, because in the early days space exploration, NASA used the most powerful computer it had: the human brain.

Kennedy addressed a joint session Congress on May 25, 1961. This was to discuss the importance the Space Race in the context the Cold War. However, the speech is still well-remembered because Kennedy stated his vision for landing a man onto the Moon before the decade was over.

I ask Congress to approve the funding necessary to fulfill the following national targets, and to go beyond the previous increases I asked for in space activities.

"First, the country should pledge to reach the goal of landing an astronaut on the Moon, and then returning him safely to Earth by the end of this decade. This period will see no single space mission more spectacular or important for long-range exploration of the universe. None will be as expensive or difficult. We propose to accelerate research and development of appropriate lunar spacecraft. We propose to design alternate liquid- and solid-fuel boosters. We propose additional funds for other engine development and for unmanned explorations--explorations which are particularly important for one purpose which this nation will never overlook: the survival of the man who first makes this daring flight. However, in real terms, it won't only be one man going on the Moon. It will be a nation if we affirm that judgment. To get him there, we must all work together.

Let it all be known that this is a judgment that the members of Congress must make. I am asking Congress to agree to a new course. It will cost the country and the Congress a lot of money: $531 million in fiscal '62. That adds up to seven to nine billion dollars annually over the next five decades. My opinion is that we should not try to get halfway or lower our sights when faced with difficulties.

This is a critical decision this country must make. I am certain that, under the leadership and guidance of the Space Committees of Congress, as well as the Appropriating Committees you will take the time to carefully examine the issue.

It is a major decision that we must make as a nation. All of us have lived through the four-years and seen the importance and the adventures in the space. Nobody can predict what the ultimate meaning is of mastering space.

I think we should go to Moon. However, every citizen of the country and all Members of Congress should take the time to consider this matter. The burden of outer space is heavy and it is not sensible to agree to or desire for the United States to adopt an affirmative posture in that area. If we're not ready to do the work and bear the burdens, we should take a decision today and every year. "

Kennedy's speech

The first issue was to find the best route to get a man to the Moon. Direct ascent to and back from the Moon was originally the plan. This was the original and best plan. Spaceships or rockets could launch from Earth to travel straight to the Moon. They would either land horizontally or in an upright position. NASA found this approach possible due to the Moon's lack of atmosphere. It was therefore relatively simple for the Moon to ascent.

James Webb was the first person to take over NASA in 1961. He inherited NASA's support and encouragement for the direct ascent plan. Wernher Van Braun, the Marshall Space Flight Center director and

William Pickering as the Jet Propulsion Laboratory chief, were the largest supporters of that plan. Mark Faget of Space Task Group was there to testify before Congress. He remembered speaking with Pickering and hearing that Pickering told him, "You don't necessarily have to go into space; . . Just aim for the Moon. When you are sufficiently close, turn on your landing rockets. . . . I thought it would make for a miserable day if the rockets wouldn't go off when you light them. "[1]

James Webb

Despite being risky the plan had plenty support outside NASA. Given the attention, direct ascent was the most likely to be chosen. LUNEX [2] was developed by the Air Force and recommended its adoption. Even though it was not adopted, development of rockets for direct ascending began. The rockets needed to land astronauts in the Moon required larger rockets than the Redstones Atlases and Atlases that the U.S. has and which were also used for Project Mercury. NASA proposed NOVA. This was a huge rocket that could take astronauts to the Moon. Although the NOVA Rockets' design was unique, it belonged to the Saturn rocket Family. However, NOVA was meant to be larger than other rocket designs. [3] NASA estimated that $7.5 Billion was required to land astronauts on Moon.

Wernher V. Braun, the Nazi rocket scientist and former Nazi rocket scientist did not fully believe in direct ascent. Therefore, he proposed another way to land Americans on Moon with the Earth Orbit Rendezvous. [4] Von Braun was an ex-nazi rocket scientist and worked with the U.S. Army for the sale of the idea. The U.S. Army claimed that Earth Orbit Rendezvous would make it possible to establish a military station on the Moon, thus capturing the strategic "highground". But Von Braun's motives were more in line with his original ambitions. He considered the Moon a step towards reaching the ultimate goal of manned landings on Mars by the year 80. Von Braun imagined spacecraft built from successive launches by rockets from Earth. In this way, smaller craft could combine to build a larger vessel. This would reduce the number of rockets required and make the plan much more plausible.

Kennedy and Von Braun 1963

The individual inspection of every component was required in order to construct a Saturn-class rocket and launch a spacecraft. Tests were performed on each machine to ensure 99.9% accuracy. If every switch, pump and light worked, Cape Canaveral might be able return to its 1,700-page management plan. 300 pounds each monomethylhydrazine and neon tetroxide, were loaded into the Command and Lunar module. At launch, the rocket and the spacecraft weighed about 6.5 millions pounds. Nelson notes that six million pounds of fuel and propellant were loaded into the Command and Lunar Modules. These included liquid oxygen, kerosene, liquid hydrogen and LH2 for stage one, as well the hypergolics [self ignition] for the smaller modules that would serve as spaceships in the final days.

Despite the fact America has a long history of rockets (such as the Redstone/Titan), there have never been any explosions at the launchpad. Because of the volume and danger involved, this is all the more amazing. Bob Jone (Kennedy rocket scientist) explained that while the LOX prechilled, with xenon lamps and wind blowing. "This thing groaning is coming on...You can see them valves and the turbo pump goin' chchchchchchchchchchch. The thing is going to scream '!"[6.

Partly due to the Saturn's successes, Apollo would take men to the Moon. The design of the rocket signaled the intersection between many ideas about American spaceflight as well as what type of agency NASA might be. The short answer: civilian. Gemini used a Titan IIA rocket, a heavy booster, as its military rocket. The U.S. Air Force also had access to the rocket for spy satellite launch missions. Gemini was also planned for use by the military branch. Geminis I to II and Big Gemini (military astronauts) would have flown them and docked them with Air Force spy station space stations. This was an example of how much the Cold War mindset had already infected much U.S. space programs. There was also

"Space Race", and there was also Rocket and Missile Race. Some saw the threat of militarization by outer space as a larger competition between the nations. [7]

Von Braun's plan had an immediate affect on the Saturn programme, as the Army Ballistic Missile Agency - closely associated with Von Braun - endorsed Saturn. This agency reported that "if a manned moon landing and return is desired prior the 1970's", the SATURN system was the only booster vehicle currently under consideration.[8] Von Braun's endorsement of Earth Orbit Rendezvous meant Saturn (and not the larger NOVA) would be the rocket to take astronauts up to the Moon. To be successful, Earth Orbit Rendezvous needed to create a system capable of launching multiple rockets in orbit around Earth and then assembling the payload spacecraft into one spacecraft system for missions into deep space.

These technologies hadn't been tested by Project Mercury. This success meant that a new project in space would have to be established. Not a linear development. Instead, it was a choice made between Apollo and Gemini. Both plans had the potential for landing on the Moon by Earth Orbit Rendezvous. [9]

Marshall team member Von Braun used Gemini as a tool to study Earth Orbit Rendezvous. A second orbital plan was also developed, which was later adopted by Apollo. Lunar Orbit Rendezvous (where spacecraft would be assembled in the Moon's orbit) was considered because there were no large rocket boosters in NASA's inventory. As a result, it was suggested in 1959 that spacecraft could be assembled directly on the Moon's surfaces. Many moon exploration ideas seemed to be based on some type of lunar assemblage. Plans were drawn up to dock and fire two spacecraft while en route to Moon. A plan was also made to send an astronaut on a one way trip, until the later rocket could reach the Moon. NASA was willing to explore other options in either case. One or more spacecrafts were to approach the Moon. With each stage, one part would be discarded until completion of the mission and crew return home.

Lunar Orbit Rendezvous wasn't an idea of one person, it was a collective effort by many people. And the idea was not taken seriously immediately. It is not surprising this, given the U.S. had yet to attempt to rendezvous any orbiting spacecraft in Earth-orbit. Vought's Astronautics Group, which had made plans for lunar-orbit in 1958, gave NASA the plan as the Manned Lunar Landing and Return. NASA had heard of the Lunar Orbit Rendezvous before, but it was not the first. NASA LOR was promoted to NASA by Dr. John C.

Houbolt of the Langley Research Center. The Apollo program benefited greatly from this idea. LOR would eventually be used to design a separate command and a lunar module. However, it was not considered serious until Von Braun and Marshall endorsed the plan. [10]

Armstrong shared a rare chance to fly with Chuck Yeager on April 24, 1963. They were assigned a T-33 aircraft and told to fly it to Smith Dry Lake in Nevada. Here they would assess the potential landing area. Armstrong later recalls, "We went to that place and looked at it. It seemed like it was dry on one side and wet on the other." Chuck asked me to do a touch & go, so I suggested that Armstrong take off and land successfully. Then, Armstrong took off again. Armstrong claimed that Yeager then said, "Let's return and try it again. And slow down even more." That worked, Armstrong says. Armstrong said, "So we tried again, and cut back the power. Then, it started to slow down. Finally, I felt it begin to soften under my wheels, so I increased the throttle. Then it settled even more and I added more throttle. Finally, we came to a stop at full throttle. It started to sink. Chuck started to chuckle. He began to laugh more. When we finally stopped, he was nearly doubled up with laughter.

Yeager

Armstrong was given the chance to join a newly formed space program not long after his incident. Armstrong later admitted that it was not an easy decision. As I was flying X-15, it occurred to me that if the project continued, I would continue as the chief pilot. I was also working to develop the Dyna-Soar. Although it was still a paper model, it was something that was feasible. The risks involved in space flight were likely to be lower than what we experienced back in Edwards flying or with the general flight test group. We were always exploring the frontiers and testing limits. This is why we had less technical insurance. We had less technical protection, less minds looking, smaller backup programs and more analysis.

Armstrong made yet another difficult landing in May 1962. He was flying an F-104 to Nevada's Delamar Dry Lake to test the area where emergency landings are possible. But as his plane came to a stop, the landing gear did not lock in place. He was able avert the landing but his ventral end got stuck in concrete. The radio was rendered useless by the release of the plane's hydraulic liquid. Armstrong was unable landing where he was. Armstrong flew to Nellis Air Force Base. Armstrong waggled at the tower and indicated that he would land radio-free. He finally touched down but his tailhook let go, catching the runway anchor and dragging the chain behind the plane. The mess was cleaned up in 30 minutes.

Milt Thompson, Edwards's agent to pick Armstrong out, encountered a severe crosswind as he was landing F-104B. This caused a puncture in his left main tires. Nellis was forced to close the runway again in order to clean up the mess. Edwards finally sent Bill Dana on a T-33 to collect the men. However, he almost landed long, convincing Nellis' powers to order ground transport to return Edwards with the men.

Armstrong found out that NASA was taking applications to pilot Project Gemini's two-man crews into orbit. They also accepted applications from civilians and not only military pilots. Armstrong visited the Seattle World's Fair and in May 1962 attended a meeting discussing the future of satellite exploration. While

Armstrong submitted his application almost a week before the June 1 deadline but Dick Day, who had learned about Armstrong and his abilities from Edwards, was able to slip it in. Armstrong passed the medical test that cleared him and Deke Day, NASA's Director in Flight Crew Operations, called Armstrong to let him know he was in.

Chapter 3: Slayton

Many people speculate on the effect Karen's death had on Armstrong's decision to leave Edwards. They also speculate on Armstrong's performance in the three months that elapsed between her death, and Armstrong's application. Christopher Columbus Kraft, Jr., at the time Director of Manned Spaceflight Operations, Mission Control in Houston, noted that "the human brain is no difference than any other computing device." It's a much better computer, but that doesn't mean it can't have its faults. The pilot won't even know it. The pilot probably does [the flying] to escape. He wants the opportunity to be back in the fray. A good flight surgeon wouldn't allow him to fly long enough. Edwards' time was when the only thing a flight physician would have done was to physically qualify him every 'x" number of days. Joe Walker, the chief flight test pilot up there should have known of the problem. He would have stopped Edwards from flying ..... Knowing Janet and Neil well, I might have not known how they were dealing with it. Both have the personality to try to ignore that.

Only the man could answer the question regarding his mental state. According to him, "I can't remember any factors from Karen losing that have influenced my writing." It was a difficult time for me personally. It may have affected me in my work, since we were going into the hospital. This is something that many families are faced with. It's not something that is unusual, but you have to deal. We did." Janet was due to give birth to Mark on April 8, 1963 in Houston.

Armstrong's early months as an astronaut were spent learning about the program. The nation also got to know him. He and the "New Nine", including Armstrong, visited sites across the country where components of the spacecrafts he would be piloting. Tom Stafford, a fellow astronaut said it was both difficult and glamorous. "We all flew commercially, four of us on one, five on the other...and everywhere where we landed we faced full schedules because we were new astronauts. But we were also supposed the men who were going on the Moon. They provided lots of food as well as plenty of alcohol. The drinking never escalated. It was a new challenge, to sign autographs, meet the chief executives and directors of major corporations.

The nine men hoping one day to reach the Moon will work hard for the next two decades and train harder than many of their military counterparts. Armstrong commented, "The first part of astronaut flight training was similar to navy flight instruction." NASA felt that the new astronauts had to be familiarized with the basics of spacecraft and orbital mechanics. I felt pretty familiar with many of these subjects. I had been studying orbital mechanics at the University of Southern California, so that was one example. Some of them were new for me, but overall the academic burden was not too much. I doubt that anyone felt the same. But it was something that we had to do."

Gemini

For the astronauts to be assembled, it was necessary to combine all the skills and talents of those involved. The U.S. space programme began with military test pilots. They then trained astronauts to fly on the moon. Rocket Men reported that Apollo 11 was over-trained. In addition to all the astronauts who spent fourteen hours in Houston simulators, Collins also had a docking simulator at Langley, Virginia, to fly; and space suits in Delaware to test. Collins was also given 10 gs of centrifuge. [12]

Apollo astronauts shared other characteristics with their predecessors such as engineering and educational backgrounds. [13] NASA hoped to train astronauts in every scenario that might arise on a mission. All astronauts who would be going to the Moon were also engineers. NASA astronauts were involved in designing, building and testing all aspects of Apollo machinery. They were experts in a particular area of spaceflight. [14] Apollo 11's first mission to land humans on the Moon was completed by each astronaut. The technical plans for the landing were also developed by them. They had shown to be the best of their field.

Buzz Aldrin, Jr., or Edwin Eugene Aldrin, was the NASA astronaut who solved the problems associated with a lunar mission. His selection to be an Apollo astronaut was more about the technical contributions he made to the rapidly growing field of spaceflight. Aldrin was also a highly decorated military man. Aldrin was an Aldrin graduate from West Point in 1951. He flew combat mission in F-86s during WWII and later worked with more jet fighters to support the U.S. Air Force. NASA finally needed his Bachelors of Engineering in Mechanical Engineering, followed by his doctorate at the Massachusetts Institute of Technology in Astronautics. Aldrin was not accepted by NASA to join the astronaut corps. However, NASA found that his dissertation entitled "Line-of–sight guidance techniques manned orbital Rendezvous", showed that Aldrin had the intellect to be an explorer. [15] He was in fact the first to get a doctorate on astronautics. This proved sufficient for the difficult task that was a lunar landing. Aldrin was selected as part of the Third Astronaut Group, despite never having been a test pilot. [16] Aldrin's dissertation demonstrates his passion. In January 1963, he dedicated his thesis to the following: "In order to make some contribution to their exploration in space, this is dedicated for the crew members this country's present and future human space programs. Only if I could have the opportunity to join them in their exciting endeavours.

Aldrin

NASA's plan for the future was changed by Aldrin being included in its astronaut corps. NASA had previously included engineers like Gus Grissom. But the inclusion Aldrin to the astronaut corps is a recognition of how huge a task NASA must complete in order to land a Moonlander. The American space agency was still undecided about the plan to reach Moon. This rendezvous could be between spacecraft orbiting Earth, or in Lunar Orbit. NASA could have taken either route in 1963. Direct Ascent still existed. NASA realized they couldn't get any orbital rendezvous between spacecraft. Aldrin, however, was chosen to make the docking possible for both craft regardless of their location.

Aldrin discovered something he called the orbital paradox while writing his astronautical thesis. It was a step forward in Aldrin's quest for a successful Moon expedition. Dr. Aldrin explains the astonishing result of this maneuver. The second craft will fly off into the distance 17, and you'll end up in an even higher orbit.

Aldrin's genius was his idea to train astronauts on how to dock with a target spacecraft using direct observation. This line-ofsight method cannot be done by one person. The pilot would utilize their direct visuals to complement the computers on board. Navigation would be a two part act between man & machine. Aldrin however felt that the combination would allow astronauts not only to navigate the craft but also make orbital rendezvous much easier. Aldrin wanted to train astronauts not to listen to their instincts as fighter pilots. Space is "a world in three dimensions," and "there was no up or down." The goal was to allow them to manually fly their craft, and dock with another, during a Moon mission.

Aldrin's plans to train an astronaut to manually pilot the craft was in line with the constant disagreements among the astronauts as well as the engineers. There will always be disagreements as to who, exactly, pilots the NASA ships -- the engineers in the ground or the astronauts in capsules. Aldrin, along with his colleagues, claimed the pilot role later in his career. It becomes obvious only in an emergency. Even in emergencies, such as Apollo 13, it was not obvious what was going wrong. It was like, 'We have a trouble, all the light are comin' on!' And it was down to the ground, to figure out what was the problem.

Aldrin did appear to stand with the engineers. However, he wasn't able to act like one during Gemini 12, which Aldrin launched into space with James Lovell on a

4-day mission to Mars in 1966. Aldrin also had to perform a 5 1/2 hour spacewalk in order to be able to manually dock Gemini with the Agena orbiting vehicle. Aldrin performed the task he originally designed without targeting. Aldrin, in some ways, was somewhere between the astronauts and engineers. Both of whom "honored," his obsession for orbital rendezvous with the "Dr. Rendezvous."[20] Nelson noted: "As an Astronaut, Aldrin exposed both his smarts as well as his ineptness in promulgating his MIT Studies so assiduously. His opinions so stubbornly that many of his coworkers were annoyed by it. Chris Kraft, flight director for the mission, stated that Aldrin's doctoral thesis concerning space rendezvous "made him, in he own eyes, one among the world's most distinguished experts." 'Dr. Rendezvous.'"[21]

Ironically Aldrin's authorship over orbiting rendezvous is what would make the Gemini Program succeed. Gemini was a mixed bag for many years, but it did the job. It showed that orbital rendezvous between two ships could be possible. This was a major accomplishment. It confirmed that orbital rendezvous in Earth orbit could be possible and was feasible. Gemini almost accomplished what was impossible by orbiting around Earth. EOR did not work for long, but Gemini was eventually viewed as the craft that would take Americans to Moon.

Gemini was called Mercury Mark II at the time of its creation, but the original identity that separated the spacecraft and Mercury was always Mercury. James Chamberlin and McDonnell Aircraft (the manufacturer of Gemini) considered Gemini to be a viable option. They believed it could do a Moon land at a fraction of the Apollo cost. They believed Gemini was capable of bringing men to the Moon within six years. Gemini would utilize the Saturn C rocket's efforts, which was considered in the 1960s as just one piece of the larger vehicle that could be used for a Direct Ascent. Gemini began its life as a docking vehicle. After that, it was transformed into a fleet with other craft that could dock to "trans-stage target vessels" (upperstage Centaur missiles) to achieve lunar flybys. Here was Gemini's schedule.

Date and Description of the Flight

Mar 1964 Gemini 1. Unmanned orbital

May 1964 Gemini 2 Manned orbital

Jun 1964 Gemini 3 7 days manned orbital

Aug 1964 Gemini-4 14-day manned spacecraft

Sep 1964 Gemini 5 Agena docking

Nov 1964 Gemini 6 Agena docking

Dec 1964 Gemini 7 Agena docking

Feb 1965 Gemini 8 Centaur docking. Lift to high Earth orbit

Mar 1965 Gemini 9 Centaur docking. Lift to high Earth orbit

May 1965 Gemini 10 LM docking

Jun 1965 Gemini 11 LM docking

Jul 1965 Gemini 12 LM docking

Sep 1965 Gemini 13 Centaur docking, boost to Lunar flyby

Oct 1965 Gemini 14 Centaur docking, boost to Lunar flyby

Launch of Saturn C-3

Nov 1965 Gemini 15 Manned Lunar orbital

Jan 1966 Gemini 16 Manned Lunar landing

Chamberlin eventually tried to combine Gemini and Apollo, in order to reap the benefits of the Saturn rocket's achievements. Gemini II was touted as a possible Direct Ascent craft. It would be a capsule that could land on Moon with an upper stage ascent module, retrograde stage and foldable landing legs. Gemini II was designed to allow the use of different modules with "advanced", Gemini's advanced technology. The goal was to have a fleet, or "Big Gemini," of craft that could rendezvous at space stations, ferry people from low to higher orbits, and even serve as a lifeboat/rescue craft for Moon explorers. Chamberlin Aircraft was convinced that Gemini was a modular craft NASA could use in Earth orbit for years. [22]

Gemini's potential is not greater than Apollo's. Apollo had a vision that allowed them to create visionary craft to carry out

LOR. NASA loved this design. It also gave them the ability to use a large booster, Saturn V (C-5), which promised to launch even more ambitious spacecraft that Gemini. The Apollo program was an unrivalled government project. It was also considered a logical follower to the Manhattan Project. Rocket Men's Nelson stated that "In so many aspects, the race toward the Moon would turn to be a sequel" to its predecessor's quest for atomic mastery. Bother were large projects that only great nations could afford to undertake and accomplish at the federal level. Both began with Third Reich immigrants and a shared geographie [New Mexico 24]

Gemini has been viewed as mostly a "trainer," for U.S. Astronauts. Therefore, the narrative of men who travelled to the Moon is clear and follows a logical path. From 1965 to 1966, nine launches followed. Every Apollo astronaut rode a Gemini-equipped spacecraft and participated an ambitious program of tests, maneuvers, evaluations, and assessments of technology. EOR, which was at the time the "cradle", of the plan for rendezvous with two spacecraft, saw Gemini's astronauts participate in those missions. The Mercury Program had inspired the Mercury Program's lessons, but the spacecraft was also constructed by astronauts with more advanced degrees. Gemini's names reflected various intentions. "Mercury Mark II", meant to signify the replacement of the first space-capsule, was a reference to its purpose. "Gus Mobile," however, was a joke about the fact that Gus Grissom Virgil Ivan Grissom who designed the spacecraft, which was for astronauts and not passengers. Gemini taught astronauts how Earth-orbit flight is possible. [25]

NASA's greatest need was a system to train and prepare astronauts for space. Armstrong was responsible for designing the flight simulators. Armstrong said that "One of my special tasks with all simulators was to check if the simulator's engineers had correctly modeled the equations." Armstrong quickly realized the difference between theory, practice, and mathematics. "I was shocked at the number of instances they weren't mechanized properly. That responsibility was natural for me because I had done the same work at Edwards; I was always making sure that the equations of motion were properly integrated into the computer." The problem was obvious: "The guys who were mechanizing the equations--sometimes contractors, sometimes NASA employees--oftentimes did not have the perspective of a pilot.

They couldn't understand what it would mean to the pilot if they were pulling up to a horizontal position, then rolling ninety-degrees and then pitching toward the ground.

NASA announced, on February 8, 1965 (through NASA's Civil Pilot Program), that Armstrong would join Pete Conrad and Gordon Cooper for Gemini 5 as the back-up crew. This mission allowed them to practice rendezvouses in space. They would also set up procedures and test equipment in preparation for a future mission. Armstrong and See saw the launch from Cape Kennedy Florida and landing at MannedSpacecraftCenter in Houston as they had planned.

Gemini 5 was Armstrong's first manned flight. NASA named Richard Gordon the pilot and Armstrong the commander of Gemini 8. Gemini's sixth and final manned flight. However, it was the first time

Gemini had docked two spacecraft. Armstrong, his pilot David R. Scott and the Gemini capsule would dock with a Gemini Agena Target Vehicle. The launch of both spacecraft was successful and the docking began without a hitch. Aldrin's "training in orbital rendezvous" program and computers helped Armstrong to pilot and dock a craft with another spacecraft.

The mission ran smoothly for 27 minutes. Scott had prepared for a two-hour spacewalk in preparation for EOR, and to train for a lunar trip. Both spacecraft began tumbling out of control. Armstrong immediately separated from the GATV. The tumbling then continued and got more intense. Gemini 8 was in control. [26] Discovery magazine published the following description of the situation: "The men from Houston got confirmation the spacecraft had broken apart as Armstrong pulled Gemini back. Scott hit the undock key. The separation didn't help the astronauts. Scott calmly said, "We have serious problem here." We're... we're just tumbling along. They were disengaged with the Agena. Armstrong said, "We are rolling up and can't turn any off." Armstrong tried unsuccessfully to dampen the spacecraft's tumbling. Scott was not able to stop the tumbling, so he had to be handed over. Gemini 8 was already making one revolution per seconds. The spacecraft was experiencing a centrifugal effect that caused loose items like checklists, procedure charts, flight plans, and checklists to stick to the walls. Their heads and arms were pinned against each other, making it difficult to reach hand controls. The sunlight reflected through the windows flashed like a strobe, and it was as bright as a flashlight. [27]

Chapter 4: Armstrong prepares for Gemini8

Gemini 8

They would be silenced by the orbit, which would cause radio silence for both astronauts shortly after they docked. Jim Lovell informed them that communications had been cut off before they lost contact. "If you have any problems and the Agena attitude control system goes wild, just send command 400 to turn it back on and take control." Granath stated, "The Agena has been designed to comply with both Gemini and ground control orders." The Agena began to use a command programme stored in its internal computer. Scott noticed that the Agena wanted to turn the two spacecraft. However, Scott was not happy with the direction they were going. He replied, "Neil.

The situation got more complicated at this point: "Armstrong used Gemini's Orbital Attitude and Manipulating System, or OAMS to stop the tumbling. Gemini VIII became out of range of ground communication and the roll quickly resumed. Armstrong began to regain control and noticed that OAMS propellant was lower than 30 percent. This indicates that a Gemini Spacecraft thruster might be the issue. Scott cycled the Agena switches repeatedly. Nothing worked. A yaw OAMS-strimmer was firing erratically. Although this was not confirmed until later it was later discovered to be caused in part by a short circuit. Crew members blamed the Agena, not knowing what it was. Scott pushed on the undock button, while Armstrong backed Gemini away the Agena. Gemini began to spin faster without the Agena's extra mass.

The men kept spinning until they were within reach of the USNS Coastal Sentry Quebec in Japan. Scott informed the ship, "We have serious difficulties here. We're tumbling end over end. We're not connected to the Agena. James Fucci from Spacecraft asked Armstrong why. Armstrong replied, "We're rolling up, and we can't turn anything on."

As the spinning speed approached 60 RPM, things got even worse for the astronauts. Armstrong considered turning off Orbit Attitude and Maneuvering System to restore control and stop spinning. Granath explained that it was not the end of his problems. He also needed to make quick decisions. Despite some problems that had occurred on U.S. spaceflights in the past, ground controllers and crews eventually found a way through them. Hodge recognized that Armstrong used almost 75% of the propellant for re-entry maneuvering to stop the spin. The crew was required to return home immediately after the RCS was activated. Hodge commanded Gemini VIII to return after one more orbit and to land within the ranges of secondary recovery forces. The U.S. Navy ship USS Leonard Mason, a destroyer of the Navy, was asked to move toward the new landing area located 620 m south of Yokosuka. ... Three U.S. Air Force pararescuers took off from an airplane to help Armstrong and Scott as they waited to be picked up by the Mason crew.

A photograph showing the crew being recovered in the Pacific Ocean

It was not surprising that the men were disappointed they couldn't complete the mission. However Scott and Armstrong won the NASA Exceptional Service Medal due to their management of NASA's first major flight crisis. Armstrong was also given a pay rise of $21,653 each year. This was more than any other NASA astronaut at that point. He then went on to serve the Gemini 11 capsule communication team on September 12, 1966. Armstrong and Janet traveled 24 days through South America in support of President Lyndon B. Johnson.

Apollo Missions

NASA started to mention LEM as its option to land human beings on the

Moon in late61. Events quickly resolved the issue of where the LEM would rendezvous and which spacecraft it would be with. NASA confirmed that LOR was ascending, and LEM development began. "Considerable research and experimentation were being conducted on engineering issues related to landing LEMs on the Moon. Dynamically scaled models were dropped on simulated lunar soil. Computer runs used mathematical models to simulate the LEM. This allowed for the study of stability as well the landing loads. An effort was also underway to deduce the lunar surface mechanics and surface characteristics in engineering terms. The engineers had to rely on the limited information available from Ranger [spacecraft]. However, later data from Surveyor and

Orbiter gave no significant improvement in the LEM design.

Apollo went at its own pace, some might call it a rush. To keep up Gemini's pace and to meet the ambitious schedule of goals the Command Service Module, CSM, and LEM were speeded along. With the advent of new technologies, the program required that its maiden flight carry three astronauts. The deadline to launch the Apollo manned flight was getting closer as every stage of Saturn rocket's Saturn rocket launched into orbit. The Apollo Spacecraft authors explained that, while Mercury and Gemini spacecrafts were being developed, the Apollo program grew in complexity and scope. The command module, service module and lunar module were all part of the Apollo program. The giant Saturn V rocket was also included. The launch vehicle and the spacecraft stood 110 meters above launching pads and weighed 3 million kilograms. The Apollo program made it possible to achieve more ambitious missions, develop more sophisticated hardware, improve the software and increase the proportion of ground support equipment.

However, the success rate was way too high. Although Soviet space advances compelled NASA to speed Apollo up, NASA's rigid schedule caused an inevitable moment. The capsule that arrived to the Kennedy Space Center launch site weighed more than 19lbs. A fire destroyed three astronauts on board, ending Apollo 1's tragic mission. After the Apollo 204 Review Board published its Final Report, they suggested two possible causes for Gus

Grissom, Ed White and Roger Chafee deaths. Apollo astronaut Frank Borman reported on the findings by the investigatory board. They believed that there was an electrical short in the lower equipment bay at Gus's left heel that led to a spark. The spark became explosive quickly after it was ignited by 100 percent oxygen with a PSI approximately twenty-one.

Apollo 1's crew

The battle to build Apollo took its toll upon engineers and mechanics. It culminated in the abandonment Launching Pad 34 and a plaque Ad Astra per aspera...A rough trail leads to stardom. NASA returned to Apollo in the knowledge that the Soviets would be taking advantage of it. Apollo 2 & 3 were cancelled. Their objectives were split into different missions and the Saturn V rocket.

Apollo 4, 5, 6 were launched in the absence of crews. Apollo received its first crew when it was considered satisfactory. NASA could replace rockets. But it had a hard time replacing highly-trained pilots. Even though they understood the "rough paths" to the "stars", it was difficult for NASA to replace them. Grissom stated that "If we die", we wanted people to accept it. We trust that if anything happens it will not affect the program. This risk is worth it."[31].

But, the space program couldn't afford to lose more astronauts during training. And this is exactly what Armstrong almost did in his preparations to pilot the LEM. NASA had an idea for a simulator to train astronauts to land at the Moon. In 1960, NASA was considering many options. NASA eventually settled on the Lunar Landing Research Vehicle. The truss made of lightweight aluminum attached to a Jet Engine. The pilot could select a mode called the lunar simulator mode. It allowed the jet engine only to support five-sixths its weight. This mode also included a cockpit which rotated freely at the engine's side, so lunar gravity could be achieved. A pilot would be seated in an open-air cockpit. He would then ascend like a flying plane, and then set the jets to behave similarly in lunar gravity. The pilot would then attempt landing the craft onto the earth using the same controls that the LEM.

Neil Armstrong was already in the running to command the first mission to the moon, and Pete Conrad was close behind. Dryden provided their training; Conrad and Armstrong respectively had made 20- and 13- flights. Armstrong's 21st trial flight, when he switched to lunar mode, went perfectly. But then the LLRV went wrong. Armstrong now flew on his left side with the simulator, and he tried to correct the trainer by using his controls. Armstrong was left without options and had to eject his LLRV. Armstrong did. He landed on a parachute at 200 feet from the ground. Armstrong's only observation to his fellow astronauts was "I had too bail out of that darn thing!"[32]. Five of seven LLRVs crashed, or exploded in flight, but there were no fatalities.

An LLRV

Armstrong's experience was a valuable learning opportunity. He was the man who commanded Apollo 11 and became the first man to land on the Moon. The military chain of command perspective, in which Armstrong was favored due to NASA's 'naval genetic code'; an engineering view, with the LEM's layout giving preference to the command plane closest to Apollo; and lastly the personality perspective. Most astronauts identified Aldrin as lacking "finesse". Armstrong was chosen because of his skills as a pilot. [33] He demonstrated the best qualities as an astronaut-engineer and was able successfully to command the Apollo spacecraft. Apollo was the epitome tech and training. Armstrong had all characteristics necessary to be able to command a mission between the Earth and the Moon, as well as make the first manned landing.

Slayton called up 18 astronauts, which included Armstrong, and said, "The men who are going to fly first lunar missions is the guy in this room." Armstrong remained steadfast as he was named to the back-up crew for Apollo 9, which was due to conduct a medium earth orbit test of both the lunar module, and the command/service module. Buzz Aldrin was the other member of the team. In the months to follow, they were swapped out with Apollo 8's prime crew. One change was that Mike Collins switched with Lovell for Armstrong's crew. Now, Lovell will fly Apollo 11.

Armstrong was part of his training to be an astronaut. He had to use a Lunar Landing Research Vehicle to practice maneuvers on the Moon's one-sixth gravity. Armstrong suffered another one of the many close calls that would define his life on May 6, 1968. The incident was described by Armstrong: "It was a typical landing trajectory during the flight that afternoon. However, as I approached landing in the final 100 feet, I noticed that my control was getting less. Soon, control became non-existent. The vehicle began to move. We did not have any secondary control system that we could start the vehicle, or an emergency system that would allow us to regain control. After the aircraft reached 30 degrees of banking, I realized that I was not going to be in a position to stop it

..... With only so much time, I decided to use the rocket-powered seat to exit the vehicle. Although I was at an altitude of approximately fifty feet, the rocket propelled me quite high. The vehicle crashed first. I was able to drift in my parachute out of the flames, and drop successfully in the middle in a patch weeds just outside Ellington Air Force Base.

Houston telegrammed NASA headquarters, Florida, that LLRV #1 CRASHED MAY 6, 1968. At 1328 CDT AT EAFB TEXAS. PILOT NEIL A. ARMSTRONG NASA.MSC ASTRONAUT. HE WAS EJECTED AFTER APPARENT LOSS OF CONTROL. ARMSTRONG SUSPECTED MINOR LACERATION IN TONGUE. VEHICLE WAS TRAINED STANDARD LUNAR LANDING. ESTIMATED ATTITUDE 200 FEET

DURING EJECTION. LLRV 1 TOTAL LOSS-- FIRSTESTIMATE $1.5 MIILLION. PROBABLE CAUSE - NOT KNOWN AT THIS TIME. PROGRAM DELIVERY PROBABLE. LLRV-2 WILL NOT COMMENCE FLIGHTSTATUS UNTIL ACCIDENTINVESTIGATION HAS BEEN COMPLETED. BOARD FOR INVESTIGATION IS DIRECTED BY MSC.

Aldrin and Armstrong not only worked on the modules but also learned how to use a lunar landing simulator. They were surrounded by dozens of observers who had been told they would be there, which made the experience less enjoyable.

Apollo 8 crashed in the Pacific Ocean safely on December 27, 1968. Lovell Borman Frank Borman William Anders were also named Time's Men to the Year.

NASA and Apollo Program were revitalized by the success story of the Apollo Program's return to orbit the Moon. NASA staff had predicted for years that the mission would succeed. Saturn could launch men into space to the Moon. Soviets had their N-1 booster and it continued to explode at the launch site or fall back towards Earth. It seemed that the U.S. was able to conquer the Soviets with the necessary space. [34] Only left was to launch Apollo 11 into orbit.

Apollo 8 takes a photo of Earth

Apollo 8's success had also benefitted its backup crew, which consisted of Buzz Aldrin and Neil Armstrong. Armstrong stated that they were "very excited about [Apollo 8]..." because there was still a problem with Saturn V and had had a few problems with Saturn V launches. Therefore, we decided to take the next launch and not just put men and crew on it but also to take it up to the Moon. We were supportive of it. We considered it a wonderful chance. If we can make it work, why not, it would be a tremendous leap forward. NASA management displayed a lot courage in making that leap. I was concerned at the time about whether our navigation was accurate enough. We could design a route that would get us around Moon at the right distance, without hitting Moon on its backside, or if we lost contact with Earth. And, if necessary, could we navigate independently using celestial navigation. We believed we could, but these were unproven skills.

Apollo 11 Final Preparations

Slayton met Armstrong in December 1968 while Apollo 8 orbited Moon. He offered to take command of Apollo 11, which was set to land on Moon. Aldrin would also serve as Lunar Module pilot and Michael Collins as Command Module Captain. NASA announced the crew in January 2009. Lovell Anders, Fred Haise, and Anders were their backup.

The announcement was widely reported and speculation quickly began about who would be first to step on the Moon. Although there have been many rumors regarding the decision over time, NASA's official history on the space program Apollo Expeditions to

Moon says, "Once it seemed quite certain that Apollo 11 had been it, newspapers and NASA officials predicted that Aldrin was going to be the first man to reach the Moon. Gemini had the EVA. Aldrin was the first to dismount. But the hatch at the LM opened on an opposite side. Aldrin had to first climb over one other bulky-suited, heavy-packing astronaut for him to emerge. That movement damaged the LM mockup and was unsuccessfully attempted. Slayton replied, "Secondly, on pure protocol grounds," that Slayton believed the commander should be the first to leave. I made the change as soon I realized that they had the timeline. Bob Gilruth approved my choice. Slayton denied that Armstrong did so. Armstrong stated, "I never was asked my opinion." Aldrin wrote, half convincingly, "It was fine with my if it was Neil."

Gemini program was ready to end its original mission of training Apollo astronauts. Apollo astronauts John W. Young & Michael Collins were on board the 10th Gemini mission. They were tasked with accomplishing the Gemini mission objectives that were left unfulfilled due to cancelled missions. Collins and Young were able to dock with Agena and fire their rockets. This enabled Apollo 8 to reach a higher orbit and rendezvous to another Agena, which was defunct and without radar. NASA was confident Apollo 8 could complete its lunar flyby after every step Gemini X took. Gemini X proved so successful, NASA awarded Collins the responsibility of command/service Module (CSM), pilot on Apollo 8.

Michael Collins

Collins' mission was nearly cancelled after an unplanned back surgery. NASA was able however to reinstate his participation in the Apollo Program and send him on the next mission to the Moon. NASA could offer Collins the seat as a pilot on Apollo 11 because of Apollo 10's successful mock-up, rehearsal and landing in lunar orbit. [36]

Collins was a rare find among NASA astronauts. He was an American Air Forces test pilot. This was very different to the larger pool NASA recruited pilots from. These typically included aviators of the Navy or Marine Corps. With over 5,000 hours of flight time, he demonstrated excellent control of "stick" and was included in Astronaut

Team Three in 1963. He was considered to be one of NASA's greatest pilots. [37] He was extremely comfortable and used to carrying out experiments. He would need to do pilot-tasks like no other aviator, and mainly the rendezvous with lunar module on-return to the Moon. Collins later commented on the test of Collins' skills as an aircraft pilot. "My secret terror for six months had been leaving them on Earth alone, returning to Earth alone. Now I am within minutes for finding out the truth. If they fail or crash back into the Moon, I am not going suicide. But if they do, I will return home. I will be a mark man for life and that is what I know."[38].

Guidance-and-Navigator system was a great way for Collins not to be "stranded" Armstrong or Aldrin on the

Moon. It was not just computers and navigation software that were important. Gyroscopes could also help pilots track the three dimensions outside space. But Collins' experience made Apollo a great success. Collins, as Apollo astronauts before and following him, would wear a patch on his eyes to alleviate the fatigue caused by endless hours of squinting. He used his sextant on the ship to view the stars, just as navigators had done for hundreds or thousands of generations onboard sailing ships. He combined the onboard data, which included the speed and position of the ship, to find the position and altitudes. This was done to start the CSM's rocket, and send Apollo 11 onto the Moon. [39]

Collins was responsible for maintaining the Guidance and Navigation unit.

Collins was responsible in setting the gyroscopes properly. The Inertial Measuring Unit (Navigator) would then compare the ship's position with Collins's, and the thrusters would be activated to steer the ship. Apollo 8 and 10 learned that the problem with this method was the inability to recognize the stars due to the sun's rays. Collins was good at using the sextant's computing power to find the stars. [40]

Apollo encountered numerous difficulties in landing men onto the Moon due to Collins' navigational skills and flight challenges. NASA faces serious gaps in information regarding the Moon's content, which could threaten the mission. Although the Surveyor and Ranger probes had done an outstanding job mapping Moon, astronauts felt that they did not have enough knowledge to assist with landing at specific locations. [41]

Collins was also worried about technology, which were responsible for the landing of the LEM Eagle at the Moon and the return trip. The LEM's LEM's upperstage, the vehicle that delivered Armstrong and Aldrin to the CSM orbit, was experiencing instability. Sometimes, it wouldn't separate from the descend stage during dress rehearsals. This would have trapped the astronauts at the Moon. [42]

Another issue that emerged was the unexpected events NASA engineers would not be able plan for. These were things they would only learn at the moment truth, when Eagle was heading to the Moon. Engineers had questions about whether imaging radar would work with the LEM, and whether the data could also be transmitted to Mission Control and CSM on Earth. NASA had to wrestle with the matter of docking Columbia/Eagle, due to NASA's past history of near disasters in rendezvous and spacewalks. EVAs from outside their spaceships often resulted in astronauts suffering from fatigue due to the pressure of maneuvering under zero gravity. NASA knew EVAs would be necessary to save the crew during an emergency. This was particularly true for Armstrong, Aldrin and others who were required to move from LEM (the CSM) to CSM. [44]

NASA flight managers as well as Apollo astronauts knew that there were many things that could go wrong and they relied upon Collins's dedication and skills to help them overcome the dangers and obstacles that could threaten the mission. His role as CSM pilot was a position of honor and responsibility. A capsule pilot must have logged previous spaceflights. NASA didn't have any Apollo astronauts in flight at that time, so Collins was chosen to fly the spacecraft around the Moon. Collins would then await the return of the two astronauts. Armstrong stated that Armstrong was conscious that this was the culmination and work of 300,000. Or 400,000 people over a period of a decade. Nation's hopes and appearances were heavily dependent on how those results turned out. And that's what allowed us to pull off this entire thing. You could walk across the street and not know when quitting was. People worked until the end.

NASA's use for civilian contractors was a mistake that caused a problem which was never fully solved by Apollo. The agency didn't have enough experience. Any astronauts who took to the Saturn rocket in a capsule were forced into inventing spacecraft systems by professionals. Despite their lack of experience, civilian engineers were able respond to the need for specific knowledge in unproven areas. They invented escape systems for astronauts in cases of launch pad accidents. [46] NASA engineers had developed many methods to rescue crew members in the wake of Apollo 1's disaster. The key system was a trio of rocket apparatuses -- the launch escape Tower -- attached to Columbia's [Apollo 11's SC] nose cone. These were ready to fire, pull them from their boosters and deploy the chutes before they sailed into an Atlantic splashdown. [47]

While engineers and technicians often found themselves in tight positions to design launch systems and other technical issues, Apollo astronauts became increasingly stressed because of doubts over the quality and duration their training. Buzz Aldrin (Apollo 11's Neil Armstrong) and Michael Collins (Michael Collins) did not feel properly trained. But they were afraid of telling [Deke] Slayton [flight operations director] this. Jan Armstrong's wife remembered that Neil used bring home his white face, so she worried about him. I was worried about them all. The worst time was in early June. Their morale plummeted. They were concerned about whether they had enough time to learn the things that they needed, and to accomplish the tasks they had to, if the mission was to succeed.

NASA also had the challenge of asking Congress for funding. They were reminded of the rocket disasters in the past and responded to NASA's request with a small budget to help fund the Saturn launcher. Many civilians involved in Saturn and Apollo thought that the Soviets prevented them creating a sufficient program to compete against them and win the race to the Moon. [49] Computers played a major role in the Apollo Program's success. It raised American confidence and trust in the ability of astronauts pilot a craft that can orbit, land, or return from the Moon. The Columbia command center had two Raytheon computers that weighed 17.5 lbs each, while the Eagle lunar module only had one. Each one had 36K memories. Although it was less impressive than 21st-century cell phones early in the 21st, each computer was more remarkable for its function and its importance. The computer system responsible to guide the spacecraft and deliver it to the target was crucial to a mission. Guidance and Navigation was tasked to guide the Apollo spacecraft across 250,000-miles of space, orbit around Moon, land at the desired location, guide the Eagle in orbit back to Columbia, and guide Columbia towards a space in Earth's atmosphere that would "capture", the capsule. Rocket Men was Nelson's description of G&N. It consisted of a miniature computer that stored an immense amount of information, an array accelerometers and geoscopes called "the inertial measuring unit" and a space sextant which allowed the navigator and the astronaut to see stars. They determined exactly the location of the spacecraft relative to Earth and Moon. They also calculated how to burn the engines in order to correct the ship's course, or land at an optimal spot on the Moon.

Computers were crucial in Apollo 11's resolution of the central problem, namely the need for a circular lunar orbit to insert the astronauts. They would be able to decide which side of Moon to go with their cooperation with G&N based on how the astronauts worked together. They could either use systems to change their trajectory to orbit or they would need to start engines to stop the mission and bring home the crew. [51] In this scenario, computers would save lives while astronauts could make the tough decision whether to land or not.

The display of the Apollo Guidance Computer located on the Columbia.

It is possible that the helpful advice given by a computer may not be followed. Armstrong and ground controls had heated discussions throughout computer simulations. NASA mission controllers worried that Armstrong wouldn't listen to computers but rely solely on his intuition. They were afraid that Armstrong would choose to go against computer simulations and disregard Houston's advice. [52] NASA was worried so much that President Nixon wrote a speech in preparation for the eventuality of Apollo 11 going down or not being able to leave the Moon. [53] Nixon said that fate had ordained that Apollo 11's men would remain on the Moon for peace. Edwin Aldrin as well as Neil Armstrong, these brave men know that there is little hope for their recovery. They also know that humanity can benefit from their sacrifice. These men will lay down their lives in the pursuit of truth and understanding, which is humanity's greatest noble goal. Their loved ones and families will mourn their loss; their nation will mourn their loss; all people will mourn their loss; the Earth, which allowed two of its sons to venture into the unknown, will also mourn their loss. They made the people around the world feel closer together through their exploration. In ancient times men saw stars and found their heroes in constellations. Modern-day men look at stars and see their heroes in the constellations. However, our heroes are epic, flesh-and blood men. Others will follow them and eventually find their way. Man will not stop his quest. These men were first, and will always remain our foremost source of inspiration. Every human being who sees the Moon in the night ahead will realize that there is an alternate world that is always for them." [54]

July 16, 1969

Apollo 11 Command Module, Service Module, and Lunar Module Design

NASA and the nation were celebrating the Apollo 8 success at the end 1967. NASA knew that the mission could not have been accomplished without a setback with a lunar module that would land on the Moon. Apollo 9 successfully returned to Earth's atmosphere after a few days of testing the lunar module maneuvers.

NASA was ready to perform an entire dry-run of landing people on the Moon, in May 1969. Apollo 10 was sent into lunar orbit, along with the command/service modules and lunar module. The first lunar module to descend toward the Moon was seen on May 22. The lunar module had been manned but the ascent station was not fully charged so it couldn't complete the entire descent. The automated system worked. Both modules were re-docked successfully. NASA was now ready for landing men on the Moon.

Apollo 10 successfully came down on May 26.

Before Collins could take the CSM with LEM attached to the Moon to pilot, the Saturn V booster must complete the most hazardous part of the mission: The launch. The astronauts rode high-speed elevators 320 feet up to the Apollo capsule. From there they were secured into their flight seats. The countdown began at two hours and 46 minutes. The Soviets attempted to build rockets for this purpose throughout the 1960s. NASA made it work with the Saturn V Rocket, which NASA still uses today as the most powerful launch vehicle. The Saturn V rockets weighed in at nearly 350 feet and carried thousands of tons. These rockets could launch a payload of over 250,000lbs into orbit. The rockets were powerful enough to lift the Apollo spacecraft into orbit for 12 minutes at approximately 18,000 miles per an hour. However, that was only the beginning for the Saturn V's duties. To escape Earth's orbit, the Saturn V had the task of accelerating the spacecraft at nearly 25,000 km/h to enable it to head towards the Moon.

This was not an easy task. NASA had also to design the Saturn V to carry out its mission in multiple stages. A spacecraft from Apollo would normally spend just a few hours in orbit about the Earth before moving to the Moon. The Saturn V's initial two stages would accelerate it to approximately 15,000mph while the final stage would take the spacecraft into Earth's orbit with a speed of about 18,000mph. The final and third stage had to also be capable of restarting the engine and accelerating the spacecraft at about 25,000 mph to propel it from Earth's atmosphere to the Moon. Saturn V was able to accomplish the job and was successful every time it used it. NASA relied on Saturn V rockets over a dozen more times than any other vehicle during the 1960s/70s without any major accidents.

After making last adjustments to leaky hydrogen fuel tanks valves, five F-1 engines powered up. The rocket lifted, moved slightly (some astronauts refer to it as a faint 'wobble), and the rockets continued their course by gimballing slightly. The thrust columns were the only thing that held the spaceship in place. The launch controller stated the words that crew members wanted to hear: "The rocket had cleared tower."

Apollo 11 was in orbit within seconds and was soon hundreds of feet high. Its auto program started a roll. The ship was on track to reach its first goal. The Saturn V rocket's initial stage would accelerate it to more than 6000 miles per hour in only two and a halb minutes. It then detachment and fallaway. The second stage, which took six minutes longer to complete, accelerated the spacecraft up to about 15,000 km/h before it crashed away. Apollo 11 will orbit the Earth in just 11 minutes if all goes according to plan. [55]

The Launch Control Center before liftoff

After two orbits on the Earth Collins was command by Houston Control, to ignite the third Saturn V stage, the "Translunar Injection". This stage would bring the spacecraft up to 25,000 mph and allow it to escape Earth's orbit. The Saturn V's third stage, which was attached to the command/service modules, disappeared on a new trajectory once the spacecraft reached escape velocity.

Apollo 11's photo of Earth in the Translunar Injection

The engines fired up and shot Apollo 11 onto the Moon. The journey would take three days, including cruising and a brief engine burst. NASA had previously mapped Moon to help select landing sites for Apollo lunar Modules. For the lunar module to be positioned for descent when the spacecraft captured the Moon, it would need to make multiple orbits around the Moon.

One Small Step

Aldrin turned on the color cam in the capsule and the historic journey began. Earthbound viewers could already see it. One thing that the camera failed to capture was the mental state within the capsule. Nelson captured the atmosphere in Rocket Men when he described Collins' feelings towards the other astronauts. Buzz Aldrin noted that Collins was the 'easy-going' guy who brought levity into everything. Collins attempted to foster camaraderie as a way to encourage camaraderie. However, this was only possible during the mission's buildup. Collins decided that Neil only transmits the surface layer of information, even though Armstrong's shield of shyness was incomprehensible. I like him. But, I don't really know what to think of him. Buzz on the other hand is more approachable. I suspect that for reasons I do not fully understand, it is me who is trying to keep Armstrong at arm's reach. I get the impression that he will probe me for any weaknesses. That makes me uncomfortable.

Perhaps, the astronauts, much like Collins, were much more comfortable in their thoughts, despite the fact that they had seen the Earth from orbit. The crew had already been inspired by the Earth's view from orbit. They began to question the "imaginary paths that you can only see".[57] After leaving Earth, however, there was no way to reference it. They were met by the blackness space; the spaceship was then enveloped by the expanses of dark and the glittering stars. [58]

At this point, all the crew could do was to go through their tasks. Collins did the task of refilling batteries, purging fuel, and disposing of the wastewater. Aldrin, Armstrong, and Armstrong went over landing procedures. They also had meals. They had to eat.

Apollo 11 astronauts were still able to enjoy the trip. Apollo 11 was able to experience a unique visual phenomenon as they traveled to the Moon. It was more difficult to see and hear the Moon the farther the ship traveled. Collins pitched Collins's ship so that crew could get a good view of their destination. Unfortunately, Collins did not have enough time to appreciate the view. They would launch the CSM single thruster on July 19, which was the fourth day of their mission. This was just as important as other events. Either

Apollo 11 was able to successfully enter the orbit around the Moon and land on its surface, or they would have to abort it.

NASA received the signal soon from the dark side Moon. Apollo 11 was now in orbit. They were "go!" to orbit the Moon. They continued to work on their mission to land on Moon. [61]

The crew looked out at the lunar surface and applauded the "perfect" orbital placement. They also exchanged notes about their perceptions on the Moon's "true" color. NASA asked them identify the "landing track" and the crew looked for identifying marks craters. A sudden panic set in when the topography almost looked identical across the board. Collins and Armstrong did eventually find their bearings. But they soon discovered that the Moon is a mysterious place. The Moon was "unwelcoming" to and "unfriendly" for the astronauts. [62] Collins later stated that the moment was so significant that "all of us know that the honeyMoon is over" and that "we are all about to put our little pink feet on the line." [63]

Collins was absolutely right. The moment came to put their "little pink bodies on the line." After 13 orbits around the Moon, Armstrong and Aldrin dressed in the EVA suits, also known as "white, thirteen-layered, Mylar-and-Teflon-coated beta-cloth Integrated Thermal Meteoroid Garments."[64] Both astronauts took their positions inside the LEM Eagle. Collins had also to be dressed up in a spacesuit to ensure he could make an emergency spacewalk.

The lunar module required both an engine, as well thrusters, to descend thousands ft to the Moon's Surface. The lunar module would descend by a pirouette in order to keep it straight. However as Apollo 11's Moon module inadvertently proved in its own way, any wrong timing of an ill-timed thrust will land a module several hundred miles away from its intended location. This can make the difference between a smooth landing, or landing among boulders, craters, or both. The lunar module was controlled almost entirely by computer, but it was manual for the final stage. It would fall off the module in the same way as the Saturn V rocket stages.

Armstrong and Aldrin did NOT sit down in their cockpits. Instead, they would fly the Eagle from an upright position with

Velcro and Tethering to the floors. LEM's appearance was more functional than beautiful, almost looking like an insect. It was made of just three aluminum foil sheets. One engineer explained that "aesthetics can be hanged." NASA's final mission to the Moon would have the bug-eyed landing craft operate as a dinghy, and it would achieve the final vision. The lower engine and stage for descent would be left at the Columbia. While the ascent and landing stages would transport the astronauts back from the Columbia, the latter would be removed and allowed to drift into space. [65]

Collins blew out the explosive bolts. The Eagle then operated under its own electrical energy. Armstrong turned the LEM 180 degrees so Collins could see the complete deployment of its landing legs. Armstrong, the mission commandant, gave the signal "The Eagle possesses wings."

This was the moment NASA, the astronauts and NASA feared most. No matter how much training, speed and orbital calculation could be tuned to a degree that was more accurate than mission planners expected. As it turned out, the first of many problems was created by the more thrust that the separation between Columbia (and Eagle) caused. It was also due to errors in calculation of gravity and its effects on the ships that the landing site for the Eagle would be four miles from the actual one, the Eagle was forced to move towards a field containing boulders. [66]

Houston was home to Mission Control (MOCR), an expansive campus-like community college. A dedicated group of controllers kept the situation under control and prided themselves in solving problems. They now would have one. The Trench was the name given to the place where the controllers were seated in front of the screen measuring 20x10 feet. It showed the two tiny dots that are Columbia and Eagle moving slowly across a large picture of the Moon. Each of their roles were designated with terms like "FLIGHT," 'CONTROL," FIDO," 'TELMU," & "GUIDO." The Trench was the place where the controllers sat. They pictured themselves as flight dynamics officers, navigation and navigators, trying to be so involved in all phases of the mission. [67]

However, none of them could see that the Eagle would miss its landing target. This was because they were too caught up in another dilemma: Mission Control had lost contact with the Eagle. Controls made several confirmations to determine whether the Eagle would continue its powered descent. The choices of "Go/NoGo" were made and the responses were unanimous. Collins's use in the CSM to relay Mission Control information to the LEM was a solution. The mission was possible to continue. Collins sent the message to Eagle, Columbia. They gave you a go at powered descent...Eagle. Aldrin replied, "We read you."[68]

Aldrin called Armstrong for the instrument readings. Armstrong was already aware that they would go over the target area almost four times.

MOCR noticed the issue as well, as the Eagle traveled faster than expected across the Moon's surface. MOCR considered aborting, but opted to ignore it. MOCR received a reply from the controller with "GOs", after which MOCR lost touch with the LEM. [69]

Computer alarms began to sound, worsening the situation. First came "Error1202." MOCR found out about the programming mistake. The alarms could have serious consequences for the mission. Both flight controllers and the Apollo astronauts were familiar with the procedure. They had been through every scenario that could go wrong with the computer simulations and they were able to use what they had learned to guide them in their next decision. "Go/No Go" Aldrin continued his descent to the Moon's Surface while

Armstrong and MOCR conferred. Aldrin would later add, "During our descent, when we started experiencing problems with the computer computer, my focus was solely inside the cockpit. I tried to relay the information from the computer to Neil so that he could use this along with his out of the window determination about where to land. Everything was moving very fast. It was simply a matter to make sure I was doing the right thing. It was impossible to reflect on the situation for too long." [71]

The next error alarm sounded, "1201.". Guidance and Navigation MOCR reflected back on the simulations training for what Armstrong thought was an inordinate amount of time. He wanted to know the reading and was worried that the powered down descent would stop. Armstrong explained later that the issue was not in the landing area but whether we could continue. As such, our focus was on clearing the program alarms and maintaining control of the machine so that it could continue flying without needing to be aborted. During this time, the majority of our attention was in the cockpit. This could explain why we couldn't see the landing site and final landing position during the final descent. It was only after we were at least two thousand feet that we could look out and see our landing area.

MOCR finally decided that the computer alarm had been caused by complex commands. It took MOCR 15 minutes to decide. MOCR confirmed to Eagle that it was able to continue the descent, despite being overwhelmed with too many commands. Armstrong maintained control of the situation and MOCR gave him a second round of "GOs." Powerful descent continued.

Armstrong placed the Eagle in the feet first position to land. He quickly realized that no suitable terrain existed for him to safely land. Armstrong flew horizontally and they flew above a craterfield with boulders. MOCR noted Armstrong's speed and the time he took to land the Eagle. Additionally, the craft was meant to land automatically. Armstrong couldn't place the Eagle at the right spot so Armstrong continued to manually control it and skimmed across the boulder field approximately 200 feet from the ground.

Armstrong had many other considerations. He knew from hundreds upon hundreds of flight simulations that a message signal would indicate a thruster shut-down, but Armstrong was aware that it wasn't the case. Therefore, he could continue landing. Fuel was an even greater concern. He now had to bring the Eagle to a stop. At 65 feet, LEM had 60 seconds left of fuel. Armstrong had instructed the Eagle to move at the same speed as MOCR. The craft entered the "deadzone" faster than MOCR likes. Even if Armstrong made the decision to abort, their descent to the surface would continue even if the ascent stage did not lift them off. They would crash, with or without fuel.

Armstrong was not able to explain why he took manual control. Aldrin never said anything. Flight controllers only heard Aldrin call for the percent fuel left and to the feet. Four forward. Drifting to the left a bit...Twenty foot, down a halb...Drifting forward just slightly; that's good 72

Armstrong saw a clearing. He began to land and the descent engines created a dust cloud. One of the Eagles' sensors had made contact. Armstrong heard Aldrin say "contact" and then shut down the engines. After relaying his famous message, Armstrong said: "Houston Tranquility Basis here. The Eagle has landed." [73]

Small miscommunications were the mark of a moment of greatness. NASA was frightened, Aldrin turned towards his religion, and Armstrong set off to explore an alien world. MOCR had a difficult decision to make after the Eagle's landing. Engineers feared an explosion in the fuel vapors but the problem solved itself. Armstrong and Aldrin were able to resume their checklists. Aldrin, who was preparing to receive Holy Communion, took his personal belongings from the Eagle as a way to get out of the Eagle. [74]

Walter Cronkite stands tall for the New York Times headline

Aldrin, Armstrong, and Aldrin were to go to bed after the Eagle landed. Aldrin was too excited to go to sleep, and Armstrong as well. Armstrong prepared for many tasks before he stepped out onto the Moon. Aldrin was also excited, so it took them 3 hours to change into spacesuits. Armstrong also had the unique task of deploying the boom arm.

It was equipped a video camera so that he could transmit images of his first moonwalk. Armstrong, who is not used to these problems, had to step 3 1/2 feet up to the Moon, but his landing proved so gentle that shock-absorbing materials did not collapse. [75]

Armstrong shot by the camera attached to the boom

Armstrong started his departure from the Eagle into the Moon's orbit just before 11:00 pm Eastern Daylight Time. Armstrong left the Eagle as he was about to leave. He activated his television camera and began a live broadcast that hundreds of millions watched. Armstrong began to tell the story of his first steps on the surface.

Historical experts have debated Armstrong's statements ever since.

They sounded like Armstrong said "That's one step for man." Others believe that Armstrong tripped over the statement. There are many theories about why Armstrong did not speak in the "central Ohioan" language. But the debate about "the missing "a" showed how many people in Earth saw his first steps onto Moon. [76]

Armstrong claimed Armstrong said that Armstrong made the remarks because he felt compelled to. Armstrong then added, "I was aware there was a chance we would be able to return safely to Earth. But the chances of us landing on the Moon are about fifty-fifty percent." Most people don't realize just how difficult the mission was. So, I felt that it was pointless to try to think up something to say if we had the need to abort landing. He said that he had once been on the surface of the water and realized the moment was right in front of him. Fortunately, I had a few hours to reflect upon it. I was of the opinion that it was a very simplistic statement. How can you describe when you get off something? It was a step. It just evolved during the time I was doing the procedures for the practice takeoffs and EVA preps. He stated, "I don't believe it was very important, but other people clearly did. Even so, it wasn't a particularly inspiring statement. It was a very simple statement.

Armstrong claimed that he meant to say "for the man" but that it was not what he actually said. People who listened to me talk for hours over radio communication tapes know how many syllables I skipped. I was not unusual to do that. I'm not very articulate. Perhaps the suppressed sound wasn't picked up on the voice microphone. It doesn't sound as if there was enough time to get the word there, as I have listened. But I am sure that people who are reasonable will understand that it wasn't an intentional statement. I also think the 'a" was intended because that's how the statement makes sense. So I would wish history would let me drop the syllable.

Armstrong let MOCR, after he'd taken his first step, know the surface and sent the message, "Yes. The surface is fine, powdery." It's easy to lift up with my bare foot. It does stick to my sole and sides of boots in fine layers. I only go in a tiny fraction of an an inch, maybe an 8th of an an inch. However, I see the footprints of my shoes and the treads of my boots in the fine, sand particles."

Armstrong claimed that, due to the lack gravity on the Moon, it was "even perhaps simpler than the simulations...It's absolutely easy to walk about." Armstrong was so exaggerated during his time on Moon that MOCR, back on Earth found out about the alarming metabolic readings.

Aldrin followed the MOCR reminder to not lock the Eagle's door. Aldrin passed that test and made his first "small step as a man". Aldrin fulfilled his desire to join the "adventures", and went one step further to another world. It only took 30 seconds or so to feel I was capable of moving around with great confidence. This is what the later crews would need. This was not just for me, it was also for their benefit. Later in the spacewalk outdoors, I did that to benefit the people back on Earth. I also did that to measure the mobility and give them an additional description. I believe that the things that were done on that first mission were made to help make subsequent missions more successful. So we'd look at the lander, take pictures, and then report back to the team. Our mission was really to run some simple experiments. The laser reflector, the passive seismicometer, to verify the leveling devices, antennas work, and then to take quick samples of the surface. Our lander, which was heavier than those of the older landers, didn't allow for enough consumables. We also didn't have enough margin to make it possible to stay on the surface for more time. Whatever the flight plan and engineers decided what our mission would look like and how many hours we could spend out there, that was it.

Aldrin's footprints in the Moon

Armstrong made it a point to obtain a Moon sample immediately to prevent the Columbia from returning to Collins. Aldrin refocused his efforts on the two tasks he had been working hard for: the collection geologically relevant samples and the use the camera to document lunar surface exploring. However, he had one problem. Aldrin only managed to capture three photographs of Armstrong on the Moon surface. The majority of those photos were not straight-on. This is due to the fact that Armstrong was only photographed three times on the Moon. They may have spent most of their time on what was probably the most difficult task of all: unfurling and pounding down the U.S. Flag to the ground. [79]

Armstrong's Aldrin-inspired picture of Armstrong reflected Aldrin himself in Aldrin's visor.

The astronauts collected thousands of rocks and samples, took seismic readings of Earth's surface, and left a memorial plaque and other tributes at the landing spot. The plaque stated that these were the first men from Earth to set foot on Mars in July 1969. They came in peace for the benefit of all mankind.

Aldrin stands in front of the Passive Seismic Experiment Package. There is a flag and an Eagle behind him.

Return to Home

Armstrong and Aldrin walked to the Ascent Step with only enough time left to get a call from President Richard M. Nixon.

"Hello Neil" and Buzz. I am calling you by telephone from Oval Room at White House. This may be the most important phone call ever made from White House.

I cannot express to you how proud my family is of your accomplishments. This must be the proudest day for every American, and I know that people from all over the globe are joining Americans in acknowledging this incredible feat.

The heavens are now part of man's reality because of your work. And as you speak from the Sea of Tranquility to us, it inspires and motivates us to keep trying to bring peace and tranquility on earth.

All people on the planet have been united for one priceless moment in history. They are one in their pride and one in prayers that you will return to safety.

Yet, there was still a problem. The ascent engine's arm switch had broken. Aldrin fixed this problem with a ballpoint pens. The Ascent Phase that engineers feared would not seperate worked perfectly as intended. The Eagle again had "wings", and the astronauts rocketed into lunar space. [80]

Armstrong and Aldrin were the first to walk on the Moon. However, Columbia was still in orbit with Collins. Collins has been called the forgotten man. For every revolution he, Columbia, made behind the far side the Moon, he truly was the "loneliest person in the world". Ironically, he belonged also to an exclusive group of people, being the only one to not witness Armstrong taking his first steps on the Moon. Collins took a lot more time alone to reflect on his surroundings, and the effect of the experience. As he saw the Earth from space again, with the sky of black and stars as his backdrop, he contemplated its condition and its future. The Earth seemed delicately balanced on its circular journey around Sun. Is the sea water pure enough to pour on your head, or do you have oily residue ? It's not easy to tell the different between a white-and blue planet and a black-andbrown one.

As he remained in orbit, his words were, "I'm alone now, truly isolated, and completely disconnected from any other life." This is what I feel... The feeling is good to me." [82]

Collins still had a lot to do in orbit and a lot to be concerned about. However, meditations and communicating with the universe are not to be dismissed. Collins was at risk when a problem arose with the Columbia's cooling system. However, he managed to resolve the problem but still had to deal with another problem. NASA had no idea where the Eagle was landed. This could have created problems for the rendezvous.

Picture of the Eagle's Ascent Stage heading towards Columbia

Armstrong in his lunar module after the landing.

Because the Moon lacks an atmosphere, it was relatively simple for the Eagle to ascend. The Ascent Stage was able to link up with the Columbia to propel it up. Once they were linked up, the Ascent Stage was launched into orbit and finally landed back on its Moon home. The ascent wasn't without its difficulties. One of the most iconic images from the Apollo 11 mission was the placing of the American flag on Moon's surface. Since the Moon is without atmosphere, there's no wind. This meant that the flag could not be folded. Since there was no wind at the surface, it was planned to make the flag last forever. The mission's planners missed one vital detail. One of the most forgotten (and least reported) aspects of Apollo 11 is the Eagle's ascent above the Moon's Moon surface. The wind generated by the Eagle's ascent caused enough wind to blow over the flag during liftoff. Aldrin watched the flag blow up and mentioned it to NASA during the ascent. NASA was made aware of this issue. He explained that the LM ascent stage had been separated. I was too busy focusing on the computers. Neil was reading the attitude indicator. Neil didn't notice me until I looked up and saw the flag drop. Therefore, astronauts on subsequent Apollo missions were careful to plant the flag further away from the lunar module in order to avoid Apollo 11's mistake.

Collins was unable even to look through his sextant, and he could not wait to see the Eagle. He finally did. After a complex series of events to get them to line up and hard dock their ports, the latches clicked into the place and the Eagle was reunited with Columbia. The hatch was unlocked and everyone began to laugh and cheer. Collins said that he remembered the moment of reunion.

The Moon orbit is only a fraction that of Earth's so the Columbia module was equipped by engines to propel itself back towards Earth. The return had to be done at an angle that allows the planet to recover it. Spacecrafts that are reentried into Earth's atmosphere will experience temperatures of around 3,000 degrees Fahrenheit. Because of this, the module has an extensive heat shield. It consists of many panels, insulation, and aluminium. Any part of the heat protection system that fails would be fatal.

Apollo 11 returned safely to Earth. On July 23, the astronauts broadcast a final broadcast and thanked all those involved. Collins informed the audience, "The Saturn V rocket, which put us into orbit, was an incredibly complicated piece, every part of which worked flawlessly... We always had confidence in this equipment working properly." All of this is possible through the blood and sweat of many people.

Aldrin added, "This has also been far more that three men on the mission to Moon; more even than the effort of a team of industry and government officials; more even than the efforts by one nation. This represents the insatiable curiosity all humanity has for exploring the unknown. Reflecting on the past days brings to mind a Psalms line. "I consider the heavens, Thou's work on them, the Moon and stars, that Thou hast ordained. What's man, Thou art mindful of him?

Armstrong said that ""The liability for this flight lies with history and with giants of science that have preceded it; next, with American people, because they have, through will, indicated their desire for it to be done; then, with four administrations, their Congresses, and finally with industry teams and agencies that built our spacecraft, Saturn, Columbia and Eagle. The little EMU was our small spacecraft on the lunar surface. We want to say a special thank you to all Americans who constructed the spacecraft. To all those people, we send a special thank-you, as well to everyone else who is listening and watching tonight. Apollo 11 would like to wish you a happy night.

Finally, it was the time for the splashdown. A number of "drogue parachute" were installed in the module to slow down fast moving objects. They could be used at around 25,000 ft. The module could travel approximately 125 MPH with these parachutes. Another set of parachutes was used to slow the module down to 20 miles an hour before the module hit the water. Once the module had sunk, divers would deploy parachutes to lift the astronauts from the water and anchor it. At the same time, a helicopter would hover above the module to collect all the items. After being picked up by divers, astronauts returned to Earth would be placed in quarantine until they could study the effects of their mission on their bodies.

Columbia's splashdown

Apollo 11 returned to Earth July 24, with a splashdown in Pacific Ocean. It would take nearly a month for the men to go through biological quarantine. NASA had devised a complex but flawed system to protect Earth from possible Moon germs, but the men returned home. Their lives were forever changed and they would all leave NASA and NASA's astronaut corps by 1970.

A photo of President Nixon visiting the quarantined NASA astronauts

Neil Armstrong

Neil Armstrong was a pioneer in walking on the Moon. In 1969, the entire world waited anxiously for him to take his first famous steps. These words were spoken by him 238,900 kilometers above the Earth.

"That's just a small step for men, but one great leap for humanity."

Learn all about the man: the pilot, the astronaut and the reluctant American hero. Learn more about his life, family, and his legacy.

Neil Armstrong on July 1969

The early years of life

Neil Armstrong was birthed in Auglaize Country on the 5th of August 1930. Viola (Stephen) and Stephen were his parents. Despite being American, some of his extended family was Scottish and Irish.

Photograph of Neil Armstrong taken as a child during the Friday 31 August 2012 memorial service.

He was the oldest of his three children. His brother and sister were June & Dean.

Stephen, Neil's dad, worked as an auditor in the Ohio state government. His job was simply to examine all the paperwork. Stephen Armstrong, his profession, required him and the Armstrongs to travel across Ohio while the children were small.

Neil lived in twenty-seven different places while growing up.

Neil fell in LOVE with flying as a child. His father took Neil to air shows at the age of two. By five years old, he had his first flight on a Ford Trimotor. It was nicknamed "Tin Goose".

EAA Ford Trimotor (NC8407).

Neil was 14 when the family moved the final time. It was back to where Neil was raised. Neil was able move into Blume High School to begin flying lessons. He was sixteen years old when his student flying certificate was issued.

He was only seventeen when he flew an airplane solo for the first time. Neil Armstrong was legally able to fly a plane on his own before he ever had a license for driving a car.

Armstrong was a Boy Scout. He wasn't just any Boy Scout. Instead, he was awarded the Eagle Scout rank by the Boy Scouts of America. He needed to earn many merit badges before he could win the award.

Neil considered Scouts to be so important that he carried his World Scout Badge to the Moon. While speaking on Apollo 11's radio station, Neil sent a message of support to his fellow Scouters.

Neil was seventeen years of age when he did very well in school. He was accepted into Purdue University, the Massachusetts Institute of Technology (MIT), and the Massachusetts Institute of Technology. He decided that he didn't want to be far from his family as they were close. He chose Purdue University.

Access to the Engineering Mall, Purdue University

Neil started college at Purdue University in Indiana when he turned seventeen. He pursued a degree in aeronautical engineers. Neil learned about the design, engineering, and technology involved when building an aircraft capable of traveling into outer space.

He was the only person in his family who went to college. American families were unable to afford college for their children at this time. But scholarships were available to talented, hardworking students who were lucky enough.

Holloway Plan provided funding for Neil's college education. This was paid for by the U.S. Navy. Those who were granted support had two years to study and three years to serve in the Navy. They would be able return to college after the service was over to complete the final two-years of their degree.

Neil Armstrong did not achieve the highest grades during his first Purdue course. Although he did not get the highest grades in his first Purdue course, he was still able to pass without much difficulty.

Navy life

Neil joined the navy when he was eighteen. For his flight training, he went to Naval Air Station Pensacola. Neil Armstrong was just twenty-one years old when he became qualified as a naval aviation pilot. He could then fly planes for navy.

He was trained to take off and land from aircraft carriers, including the USS Cabot & USS Wright. These U.S. Navy ships were specifically built to carry planes around the world.

Once he was qualified, he was assigned as a pilot to the Fleet Aircraft Service

Squadron 7 San Diego. His second assignment involved him flying an F9F-2B Panther.

Two U.S. Navy Grumman F9F-2B Panther fighter jets over Korea, 1951-52. Neil Armstrong was the pilot.

1951 began with the landing of a plane onto a carrier. It was on the USS Essex.

Neil was just twenty one years old when he was promoted from Midshipman to Ensign that same year. An ensign, or junior officer in navy, is an individual who has been promoted from midshipman to ensign. He was promoted after demonstrating skill and dedication.

Armstrong was an officer in the navy during World War II, which took place between 1950-1953. This was a conflict between North Korea and South Korea. America and United Nations supported South Korea.

U.S. aircraft carrier USS Essex.

Armstrong experienced first warfare in August 1951. He was part in reconnaissance missions. This is where the armed services investigate their enemy territory. These missions can be very dangerous and Neil's was no different.

Armstrong crashed his plane during this period. He was struck by antiaircraft fireworks. After hitting a pole six meters above ground, he was able to rip off one meter from the wing. He was able to fly the plane safely back into friendly skies thanks to his skills. He was forced to jump from the jet after it reached friendly airspace because he knew the plane was going to crash. He would have most likely died if it had hit him. He was saved by his jeep driver and fully recovered.

Neil flew seventy-eight flights over a short period in Korea.

Neil received the Air Medal for his initial twenty missions, followed by a Gold Star award for the twenty following. With seventy eighth missions, he continued receiving Navy recognition, including a Korean Service Medal and an Engagement Star.

Armstrong was discharged from the Navy in August 1952. He wasn't yet done serving the Navy, but he had signed up to join the U.S. Naval Reserve, as a junior grade lieutenant. Reserves are part-time. They must train and are paid basic pay. They are available in case of conflict. Neil remained in reserve for eight years.

Neil Armstrong also worked in research pilot. He also flew planes still in development. F-100 Super Sabre A/C and F-101 Voodoo variants were also flown by him. Lockheed F-104A Starfighter is another of his aircraft. Bell X-1B, Bell X-5, North American X-15, F-105 Thunderchief, F-106 Delta Dart, B-47 Stratojet, and KC-135 Stratotanker.

An American Air Force Lockheed F-104A-10-LO Starfighter.

Returning to College

Neil left the Navy to go back to college and found a new passion and focus. This led to him achieving higher grades. His return saw him become a highly active and involved member of the university's many clubs, groups, and committees. He was popular with many of his friends.

He was accepted into Phi Delta Theta fraternity. A fraternity, or social group of students, is a type of student organization. Each group has a history and tradition. Phi Delta Theta was no exception. It was formed in 1848. You have to make a promise and take a pledge to be a part of the fraternity.

Neil was an accomplished musician. He played the baritone (a type of brass instrument) in National Honorary Band Fraternity. He was part the Purdue All American Band. A Kappa Kappa Psi fraternity organization, he was also a part of. Kappa Kappa Psi (his main fraternity, Phi Delta Theta), was rich in history, having been founded in 1919.

Kappa Kappa Psi was a group that offered opportunities to its members to gain leadership experience. This group also included Bill Clinton (former president of the USA).

Neil had the chance to exercise leadership skills and express his creativity by co-directing his own student musical. The musical was a great success.

Neil Armstrong graduated from the University of Arizona in 1955 with a Bachelor of Science degree in Aeronautical Engineering. It had taken him eight long years. He had also spent some time in active combat, something that is rare among students.

Armstrong returned later in his career to get his masters degree in aeronautical Engineering from the university. A master's is more difficult than a first degree student. The work is at an even higher level. He chose the University of Southern California as his university.

Neil Armstrong's statue at the Neil Armstrong Hall of Engineering Campus, Purdue University West Lafayette.

Love, marriage, and families

Janet Elizabeth Shearon became his wife during his second Purdue stint. She was studying domestic economics. The couple never considered their relationship to be a great love story. Janet was never officially asked to marry Neil.

Janet and he were married in Wilmette Illinois on 28th January 1956. Neil was twenty-six, when they married. Their married life began with them living apart. Janet was in Los Angeles and Neil was at the airbase. But, this separation was not permanent. Within a few weeks, they bought their first house at Antelope Valley in California.

Janet and Neil have three children: Eric. Karen. Mark. Karen suffered from a brain tumor that left her unable to walk, talk or even speak. Karen was two years old when she died in 1962.

Janet gave her all to her husband. She didn't complete her home economists degree. She regretted this later in her life. After 38 years together, they were divorced in 1994.

Neil married Carol Held Knight, his second wife, shortly after Janet had filed for divorce. They lived together in Ohio on a farm.

Neil's Initial Job

Neil left Purdue to become a pilot and then began his first career at Lewis Flight Propulsion Laboratory. This job involved flying experimental planes and testing their performance.

But he didn't remain there very long. He began a new job with Edwards Air Force Base, as a testpilot, in July of that same year.

Neil Armstrong while he was a testpilot at Edwards Air Force Base. 20 novembre 1956

Armstrong's job title was experimental researcher pilot. His first duties included flying chase planes. A chase airplane is one that follows another aircraft to help it monitor, photograph or give guidance and support when the plane is in trouble. Armstrong flew alongside experimental fighter planes.

Armstrong had to handle his first flight incident in March 1956 as a test pilot. He was piloting a Boeing B29 Superfortress. After they reached 30,000ft, the number 4 engine gave up. Neil was forced by circumstances to use just one engine on landing.

Armstrong flew a rocket plane his first time in 1957. The Bell X 1B rocket was used. The landing gear failed during the first flight. He flew many test runs in rockets, mainly in the North American X-15. He flew to 207 miles and 500 feet.

Neil Armstrong, a Dryden pilot and photographer, is seen here after a 1960 research flight.

Numerous incidents happened while Neil was at Edwards. They have since been immortalized. He flew planes that had their noses facing upwards, he overshot the landing strip, and even got a plane stuck with mud.

Neil was at Edwards when he was involved in a famous incident called the Nellis matter. Armstrong made a mistake. Armstrong made a mistake when he was landing. He misjudged which angle to use and lost his radio and landing gear. He had no choice but to fly to Nellis Air Force Base for an emergency landing. Another plane was dispatched to pick him at Nellis. It ran into trouble again and had emergency landing. It was decided that everyone should drive back to Edwards instead of risking flying for the rest.

Neil Armstrong got to exhibit his engineering skills while he was working as a testpilot at Edwards. He was well-known because he could quickly pick up new skills.

Armstrong reached Mach 5.74 in x15-1.

Neil Armstrong, after a 1969 research flight, in the cockpit the X-15 (a rocket engine-powered aircraft).

Neil flew over 200 different aircrafts during his time in test piloting and logged over 2,400 hours.

Neil Armstrong: The beginning of an Astronaut career

Neil Armstrong was invited to join the U.S.'s "Man in Space Soonest" program. This was a project that was established to attempt to defeat the Soviet Union, and to create the first human in space.

NASA's Project Mercury was later renamed. Neil was not the only person who made it into space.

He was chosen to be a pilot consultant for the X-20 Dyna Soar space plane. This was being designed by Boeing for use in the U.S. Air Force. Neil was one the seven pilot engineers that were selected in advance to fly a newly developed spaceplane, even though it was still being designed.

NASA (National Aeronautics and Space Administration), is the organization that was behind the Apollo programme. It was heavily funded and set out to land a human on the Moon by the close of the 1960s. Armstrong was one the pioneers in this field.

Neil Armstrong was asked to officially join the NASA Astronaut Corps on 13

September 1962. The media was as enthusiastic as Armstrong; they called the nine selected pilots the New Nine. Armstrong and his colleagues were big news.

Manned Spacecraft Center of National Aeronautics and Space Administration in Houston has selected nine new flight crew members Sept 17, 1962. Neil Armstrong, second from left, is on the first row.

Armstrong was selected as one of the two civilian pilots. A civilian refers to someone who isn't a member of the armed forces. The other civilian tragically died in a crash of a plane before he had the chance to go into space. Armstrong was, therefore: the first American civilian who reached space.

Gemini

Gemini 8 was Neil's first space venture. Armstrong and David Scott were announced to have flown the seventyfive-hour mission as preparation for going to the Moon.

Portrait of Gemini 8 Prime and Backup Crews. Gemini 8's primary crew is composed of pilot David R. Scott and command pilot Neil A. Armstrong. Standing in support crew are astronauts Richard F. Gordon Jr., command pilot, and Charles Conrad Jr. 4 November 1965

Gemini was established to develop and practice the skill of orbital space rendezvous (or docking) and practice it. This is the process by which two spacecrafts connect while in orbit. In order to fly successful missions to the

Moon, astronauts would need to feel comfortable doing this for the future Apollo program.

Armstrong is shown in the final stages of preparing for Gemini's launch.

On March 16, 1966, the mission was officially launched. Scott and Armstrong were flying in an aircraft called Titan II. They were following the Agena, an unmanned satellite craft. The mission was a rendezvous with Agena, and to practice docking with it. Armstrong and his copilot were able successfully to dock with the Agena and rendezvous within 6.5 hour. This was a record.

Gemini 8 launched from Kennedy Space Center, March 16, 1966 at 11:41 A.M.

However, the mission did not go smoothly. They kept losing radio contact to ground control. Safety is important when space travel takes you out of touch with Earth. Armstrong and Scott had a hard time docking the spacecraft because the radio was out of service.

After the docking, they began to roll. Scott and Armstrong spun in space. Scott had to make several attempts to stop this spinning. When they made contact with Earth, ground controller advised them to undock. However, the spinning kept getting worse. They were turning once per second!

Scott and Armstrong, who had no control over ground control, turned off OAMS (Orbital Attitude and Maneuvering System) on the Titan. It was believed that it was an issue within their own spacecraft. They were later proven wrong. The Titan II had faulty electric wires. They had to return from the space mission incompletely.

David Scott (left) and Neil Armstrong (right), open their spacecraft hatches while waiting for the arrival the recovery ship after they have completed their Gemini VIII mission.

Some astronauts office personnel heavily criticized Scott, Armstrong, and others for not following correct procedure. Walter Cunningham went public with his anger. He felt the mission could be completed had the pilots made different decisions.

Walter made false claims. There were no malfunction protocols. Gene Kranz who was the investigator into the incident stated that "...the personnel acted as instructed, but they reacted in a wrong way because they were incorrectly trained.

This near catastrophe taught the space missions a valuable lesson. Two spacecraft must be treated as one unit when they dock together. These lessons allowed for a successful Moon landing.

Neil was affected by the event. He was disappointed to learn that the mission failed to achieve many of its objectives. This included the opportunity for his fellow pilot, Neil, to walk around the spacecraft in space.

Gemini 11 is the last of Gemini's projects. Armstrong, who was part of the backup squadron, was so well-trained in spacecraft systems that he was often viewed as a teacher to William Anders.

Gemini 11 went into orbit on September 12, 1966. The mission was complete success.

The Apollo Program

The first Apollo missions ended in disaster. It caught fire on the 27th of January 1967, killing three astronauts. It wasn't even in space at this time. The pilots were just training. The crew became trapped when the cockpit caught on fire. It was a reminder for astronauts the dangers they were up to.

Portrait of Apollo 1's Apollo 1 prime crew, for first manned Apollo orbit flight. From left, Edward H. White II, Virgil I. Roger B. Chaffee. "Gus", Grissom. All three men lost their lives in the cockpit fires of 1967.

Apollo's mission didn't end with this sad incident. Armstrong and seventeen other Apollo astronauts heard in April the same year that there was a good chance they would fly to the Moon.

Armstrong was scheduled to be part the Apollo 9 Mission, which was to carry out a medium Earth orbit test to the lunar command center service. This would be an evaluation of the spacecraft which would be used in landing astronauts on Moon.

Neil Armstrong now commands Apollo 11, the spacecraft that was built for the Moon.

Apollo spacecraft (SM=service module, CM=command module, LM=lunar module).

Armstrong was able to continue his intense training. To be able handle extreme conditions, astronauts have to be fit. NASA made the conditions for his work to be able to reproduce the Moon's atmosphere.

The Moon is one sixth as gravity as the earth. This means your feet wouldn't stay on Earth if you were on Moon.

Neil's controls became less responsive while he was working in a replica spacecraft. He was at 100 feet above ground. He had to expel and did so safely. However, he did bit his tongue very badly. This incident could have easily killed him.

Neil Armstrong parachutes down at Ellington Air Force Base (6 May 1968) as Lunar Landing Research Vehicles (LLRV) no.1 collides with it.

Neil insisted that it was important to train in replica vehicles even though the training was difficult. Armstrong believed that these replicas were crucial if the astronauts ever want to succeed on landing on the Moon.

Apollo 11

Apollo 11 was called the mission that flew on the Moon.

It was manned by Neil Armstrong, Buzz Aldrin (and Michael Collins) and launched on July 20, 1969.

Apollo 11 crew. To the right Neil Armstrong (left), Michael Collins (right), and Buzz Aldrin.

The mission was to return to Earth from the Moon safely. The event aired live on television, and was also broadcast to radio stations around the globe.

Kennedy Space Center, Kennedy Space Center.

The Apollo spacecraft flew quieter and was larger than other spacecrafts Armstrong has flown in. Neil was previously afflicted by motion sickness. However, the flight was much more pleasant and smoother than previous flights, so he had no symptoms this time.

Photographed from the Apollo 11 satellite (16 July 1969): View of Earth

The Eagle is the name of the remote vehicle Buzz Armstrong used to land on Moon.

Armstrong reported that Armstrong was going too fast as the Eagle approached the Moon. Warning lights began flashing, alarms also went off.

Armstrong and Aldrin were unaware of the problem. They were assisted in Houston by ground control, who gave them instructions for how to fix their problem.

Armstrong saw as they were landing that the area was unsafe. He had to switch between manual and automatic controls to safely land. Although he was trained and certified to manually land the Eagle it was not the right way. It took much longer than expected.

Apollo 11 Lunar Module, The Eagle, photographed in lunar orbit, from the Command and Service Module Central (20 July 1969).

There was concern that the landing took too long and that they wouldn't have enough fuel in order to get back. Buzz and Neil thought they had less fuel than twenty seconds in the Eagle. If that were the case, they would still have to stay on the Moon!

Buzz Aldrin (Nathan Armstrong) and Neil Armstrong touched down on the Moon July 20, 1969 at 8.20 PM.

Armstrong said that the eagle had landed. Buzz and Neil exchanged handshakes. Ground control said, "We copy what you do on the ground. It looks like a lot of guys are about turning blue. We're feeling better. Thanks a lot."

Apollo 11: Landing site at Sea of Tranquility

Once they were finished, they depressurized Eagle's engine, opened the hatch and descended down the ladder.

Neil Armstrong was the pioneer. Neil Armstrong took the first step on the Moon at 2.56am, July 21, 1969. He said that "That was one small step by a man, one huge leap for mankind." More than 450,000,000 people have heard it.

Neil Armstrong descends the ladder in the Apollo Lunar Module for his first step on The Moon.

Buzz joined Neil twenty minutes after, and the pair began their work on investigating the Moon's surface. They put up an American flag, unveiled plaques, took pictures and even talked to Richard Nixon, the president of America. Armstrong also left an American and Russian memorial package to honor all those who perished in the race toward the Moon.

Image of Buzz Aldrin in the Moon. Neil A. Armstrong's photo shows the Lunar Module (Eagle), as background.

Although the Moon walk lasted for just over two hour, NASA was unsure about how the space suits would perform in the moon's atmosphere. Buzz and Neil's experience has proven how durable space suits really can be.

With Neil Armstrong, Buzz Aldrin, the Astronaut on the Moon.

The mission wasn't over yet. They had to be home safe first. Buzz and Neil broke the ignition switch of one of their engines as they got back into Eagle. It was because their space suits had been so large and bulky. They had the task of dismantling a pen in order to replace it.

Apollo 11 lunar module Eagle returned from the moon surface to dock alongside the command module Columbia.

They successfully completed the mission and returned to Earth well-rested. They re-entered into the atmosphere and landed at the Pacific Ocean. The celebrations had a wait as they had 18 days of quarantine to check for any infectious diseases or diseases that might have been transmitted from the Moon.

Apollo 11 astronauts in Biohazard Suits after Recovery from the Pacific.

Life after The Moon

Armstrong's last mission into space was the Moon Landing. He started a new position in the Office of Advanced

Research and Technology of Advanced Research Projects Agency. After only one year, he decided to leave NASA as an employee full-time.

Valentina Tereshkova presented a badge for Neil Armstrong in commemoration of his visit (1 June 1970) to the Gagarin Cosmonaut Training Center.

Neil began his career as a teacher in Aerospace Engineering at University of Cincinnati. He retired eight years later.

NASA was not his final assignment. His advanced skills and knowledge in space flight allowed him to continue his work. He was involved with the Apollo 13 & Challenger disaster investigations.

Neil Armstrong was also an ambassador for Chrysler and other businesses. He could make use of his fame to get companies the attention they desired. His connections to business extended beyond his personal life. He was a member of the board of directors at other companies such as Marathon Oil or United Airlines.

His varied career after being astronaut also included voice acting. He was the voice of Dr. Jack Marrow, in Quantum Quest. NASA partially funded this Sci-Fi adventure.

Neil Armstrong's Scottish roots are traced back to Langholm in Scotland. In 1972, Armstrong stated that he considered this tiny Scottish town to be his home. Neil couldn't go back to this small town because of an outdated 400-year old local law. This said that any Armstrong family member living in the community would be hanged.

Neil and Carol shared a farm. The farm was situated in Ohio. Neil enjoyed the fresh air and worked on the land.

Neil Armstrong receives United States President Jimmy Carter's First Congressional Space Medal of Honor (1978).

Neil was in a tragic farming accident in 1979 when he lost the tip to his finger. His wedding band got stuck in the wheels of a grain truck. He saved the finger by putting it in ice immediately and taking it with him to the hospital. His fingers were saved by the surgeons who were able make him whole again.

1991 was a difficult years for Neil. He was still mourning his parents' deaths and suffered a slight heart attack in Aspen while skiing. He lived for another twenty-one year, but he was able to recover.

Neil Armstrong spoke at a dinner celebrating the 30th anniversary Apollo 11 Moon landing.

Neil Armstrong's death

Neil Armstrong died 25 August 2012 in Cincinnati. He was 82 and died from complications after undergoing heart bypass surgery.

The White House immediately released statements claiming that Neil was "among American heros - not just in his time, but throughout history... Armstrong had carried American citizens aspirations and had delivered."

Neil Armstrong getting out from the cockpit of an Ames Bell X-14.

His family said Neil was a "reluctant American patriot (who had), served his nation proudly as a navy fighter pilot test pilot and astronaut. While we mourn his loss, we also celebrate and hope that his extraordinary life will serve as an example to young people across the world. We remember him for being a good man and we miss him. There is one thing you can do to honor Neil. Remember his modesty, dedication, and service. Next time you go outside and see the Moon, remember Neil and give him one more wink.

This request now has a legacy on Twitter with #WinkAtTheMoon

Apollo 11 crew quarantined following the moon landing. Neil Armstrong is at the left, having a good time.

Buzz Aldrin said that he was "deeply disturbed" by his loss. I know I am not alone in grieving the passing a true American hero, and the best pilot I ever met." NASA administrator Charles Bolde said that Neil Armstrong would be remembered "as long as there is history books, Neil Armstrong" for his first small step in a universe beyond our own.

Neil Armstrong was cremated. His ashes were then scattered at sea in USS Philippine Sea. Washington National Cathedral paid tribute to him. Flags were flown in half-mast to honor him.

During Neil Armstrong's burial at sea, the U.S. Navy ceremonial crew holds an American flag in front of the ashes.

Legacy

Armstrong was often referred to as an American hero. But he declined the honor. His friends described him as a humble person.

Buzz Aldrin took a photo of Neil Armstrong, shortly after he completed a lunar walk.

www.ingramcontent.com/pod-product-compliance
Lightning Source LLC
Chambersburg PA
CBHW050407120526
44590CB00015B/1860